Encyclopædia Britannica, Inc., is a leader in reference and education publishing whose products can be found in many media, from the Internet to mobile phones to books. A pioneer in electronic publishing since the early 1980s, Britannica launched the first encyclopedia on the Internet in 1994. It also continues to publish and revise its famed print set, first released in Edinburgh, Scotland, in 1768. Encyclopædia Britannica's contributors include many of the greatest writers and scholars in the world, and more than 110 Nobel Prize winners have written for Britannica. A professional editorial staff ensures that Britannica's content is clear, current, and correct. This book is principally based on content from the encyclopedia and its contributors.

Introducer

Steve Jones is professor of genetics and head of the biology department at University College London and president of the Galton Institute. He is one of the most acclaimed writers and thinkers about genetics in the world. He appears on television and radio regularly, and writes in numerous newspapers and journals. He is the author of several books, including *The Language of the Genes, In the Blood, Almost Like a Whale*, and *Darwin's Island.*

Also available

The Britannica Guide to the 100 Most Influential Americans
The Britannica Guide to the 100 Most Influential Scientists
The Britannica Guide to the Brain
The Britannica Guide to Climate Change
The Britannica Guide to the Ideas
that Made the Modern World
The Britannica Guide to India
The Britannica Guide to the Islamic World
The Britannica Guide to Modern China
The Britannica Guide to Russia

GENETICS

The most exciting developments in life sciences – from Mendel to the Human Genome Project

Introduction by Steve Jones

RUNNING PRESS
PHILADELPHIA · LONDON

Constable & Robinson Ltd
3 The Lanchesters
162 Fulham Palace Road
London W6 9ER
www.constablerobinson.com

Encyclopædia Britannica, Inc.
www.britannica.com

First published in the UK by Robinson,
an imprint of Constable & Robinson, 2009

Compiled by Anne Waddingham

A copy of the British Library Cataloguing in Publication
Data is available from the British Library

UK ISBN 978-1-84529-944-6

1 3 5 7 9 10 8 6 4 2

First published in the United States in 2009 by Running Press Book Publishers

9 8 7 6 5 4 3 2 1
Digit on the right indicates the number of this printing

US Library of Congress Control Number: 2008931537
US ISBN 978-0-7624-3620-0

Running Press Book Publishers
2300 Chestnut Street
Philadelphia, PA 19103-4371

www.runningpress.com

Printed and bound in the EU

CONTENTS

List of Illustrations vii
Introduction ix

Part 1 Building Life
1 Molecules of Life 3
2 Understanding DNA and Genes 34

Part 2 Genetics Through Time
3 Historical Background 53
4 Heredity and Evolution 80
5 Evidence for Evolution 124

Part 3 Genetics and Human Health
6 Human Genetics 217
7 Human Genetic Disease 234

Part 4 Genetics Today and Tomorrow
8 Animal and Plant Breeding 275
9 Biotechnology 303

Part 5 Controversies in Genetics
10 Ethical Issues in Genetics 331

Glossary 367
Index 371

LIST OF ILLUSTRATIONS

The double helix structure of DNA 6
Protein synthesis 15
Mendel's law of segregation 84
Mendel's law of independent assortment 86
Meiosis 102
DNA phylogeny 120
Effect of nucleotide substitutions on codons for amino acids 137
The Harvey–Weinberg law applied to two alleles 143
Adaptive radiation: 14 species of Galapagos finch that have evolved from a common ancestor 180
Genetic differentiation between populations of *Drosophila willistoni* 186
Sketch of *Anomalocaris canadensis* 202
Phylogeny based on nucleotide differences for cytochrome c 206
Pedigree of a family with a history of achondroplasia, an autosomal dominantly inherited disease 242

Pedigree of a family in which the gene
for phenylketonuria is segregating 247
Pedigree of a family with a history of haemophilia A,
a sex-linked recessively inherited disease 248
Steps involved in the engineering
of a recombinant DNA molecule 310
Pedigree of a family with a history
of Duchenne muscular dystrophy 336

INTRODUCTION

Steve Jones

Genetics has been hailed as the science of the twenty-first century. It received a similar accolade in the twentieth (DNA and all that) but in fact its questions asked go back far further, to the Book of Genesis – the first genetics text of all – which, like this *Britannica Guide*, asks some remarkably up-to-date questions. Where do we come from? How did life begin? Why are there two sexes? Is our fate inborn or is it shaped by how we live? When and why will I die? With whom among my relatives am I allowed to mate? Who belongs to our family, to our tribe, and to our nation? Is man on a unique and divinely inspired pinnacle or is he just an elevated ape? The Book of Genesis even has an early insight into cloning, with Eve constituted from Adam's rib (inbreeding creeps in too, with the incestuous parenthood of Cain and Abel).

To some people, genetics itself has become a sort of religion. It is a matter of faith; a curse or a salvation, promising or threatening its believers according to taste. People expect a lot from DNA: doom or salvation, heaven or hell. In fact, the study of inheritance – as this book makes clear – is most of all a science, and in places a rather complicated one. It tells us almost nothing about human affairs that we did not know before, but it says a lot about how we work, where we came from, and even a little about where we may be going.

The issues raised may be ancient, but the answers we have so far, such as they are, were a long time coming. Genetics is, almost, a science without a history. Unlike physics, which can be traced back to the Greeks, or chemistry, which descends from the alchemists (both of whom knew a lot more than they are given credit for), it traces back to just one man, Gregor Mendel, whose paper on peas, published in 1866, was promptly lost from view.

Plenty of people before the famous abbot, and many more in the forty years that his work stayed in obscurity, tried to understand how information is passed from one generation to the next; but without exception they were barking up the wrong tree. Darwin believed in the inheritance of acquired characters. In that sense he, with most of his contemporaries, was a Lamarckian (the much-quoted disagreement with the French biologist turned not on the mechanism of inheritance but on the latter's idea of a sort of inevitable progress in evolution, a notion that Darwin despised). He then turned to an idea based on a sort of blending of the blood, but that was disproved by Francis Galton who transfused blood between silver-grey rabbits and those of normal colour to no avail. His cousin stayed puzzled about the transmission of information from one generation to the next (and he worried for years about the damage that the diluting-out of favoured characters caused by blending inheritance did to his theory of evolution).

In fact Charles Darwin, and a few others, almost got there. In his book on the domestication of plants and animals, he discusses round and wrinkled peas, and in one of his thousands of letters even describes a human attribute – an inherited absence of sweat glands – which appears almost exclusively in males, but is passed through females (a perfect example of what is now called sex-linked inheritance). Lots of other biologists, too, noticed "atavisms" and "throwbacks", the sudden appearance of attributes present in distant ancestors (and due to the inheritance of rare recessive variants in the double copy needed to manifest their effects).

However close they came to the truth, Mendel was the only one to get the answer. The pea work was rediscovered in the early years of the twentieth century and the science of inheritance entered a phase of explosive growth. As a result, geneticists of a certain age (and I count myself among that motley crew) are in the situation of a chemist meeting Mendeleyev or a physicist introduced to Isaac Newton: for the founders of our science lived on into own time. Forty and more years ago I was taught introductory Mendelism by Lotte Auerbach, who discovered how chemicals can damage DNA and collected fruit flies (also known as vinegar flies) with Theodosius Dobzhansky, one of the founders of experimental evolution. Until the middle of the century, DNA was still widely dismissed as "the stupid molecule" because it was so chemically simple – but James Watson, half of the pair who killed that adjective for ever, is still magnificently (and controversially) around and Sydney Brenner, who opened the door to understanding how simple chemistry can be translated into complex biology, is, when not on an aeroplane, happy to puncture the scientific ideas of anyone daring enough to dispute with him.

The Britannica Guide to Genetics is the third scientific work in this distinguished series, the first being *The Brain*. The second book deals with the other pressing question of the twenty-first century, climate change. It is no coincidence that it should so soon be followed by a volume on the science of inheritance, which alarms at least some among the public with what might be threats even more dire than global warming. Eugenics, genetic decay, genetically modified (GM) crops, designer babies – all, to some, menace the future. As is true for the greenhouse effect, an objective account of the science behind the hysteria is what is needed, and this guide sets out to provide it. Its subtitle – *From Mendel to the Human Genome Project* – marks the beginning and the end of a period in genetics that may turn out to be just the overture to a long and challenging series of later acts.

Confident as geneticists may have been in my days as an undergraduate, their subject had, as we can read in the following pages, scarcely begun. The double helix, which became the international icon of fate and the future, was of interest only to biologists; and the notion that it might one day be read from end to end seemed a fantasy. Genes were still, more or less, beads on a string; and the dogma was of one gene, one enzyme; and one enzyme, more or less, one physical attribute. Genetics was still a sexy science, at least in the sense that most of it turned on breeding experiments. Many were ingenious indeed, but at best they gave an indirect insight into the machinery of inheritance. The textbooks on the subject were devoted to the stern truths of the fruit fly and the fungus and their account of human inheritance came, invariably, as a last (and often skimpy) chapter, sometimes with a compulsory nod at the horrors of the eugenics movement.

Mendel succeeded because he was, more or less, the first biologist to count anything. Physics and chemistry have already become the "mathematization of knowledge": the replacement of vague generalizations with a precise arithmetic understanding of at least part of the world around us. Biology stayed (and in some ways still stays) oddly insulated from that stern truth. As this guide shows, genetics forced the science of life to become more serious and, since then, geneticists and many of their fellows have been dragged (more or less reluctantly) into the world of mathematics. Statistics itself began with the study of inheritance, and many of its techniques were developed to test whether the ratios of particular types in succeeding generations fitted with those expected on Mendelian theory.

This volume explains his basic rules, now much modified, and shows how the double helix complements earlier experiments on the genetic material. Our own DNA sequence, the infamous "book of life" with its three billion letters was read off a few years ago and, unsurprisingly, is full of surprises. It structure resembles that of one of the many fictional blockbusters that have brought some version of genetics to the public eye: rambling, badly constructed, with lots of repeats in which the same idea comes up again and again with minor variations. There is also lots of inserted "gunk" which is edited out and thrown away, with no apparent loss of quality (although quite a lot of garbage is left). Now and again there is a burst of repetition, a gap in the narrative, or a sudden and unexplained shift from one subject to another. Overall there seems to be little logic in how the genome is put together. The most unexpected discovery was the apparent shortage of functional genes – not hundreds of thousands, as expected by some, but about 25,000 (which may tell us more about how little we know about genes than about what lies behind human inheritance). The genes themselves have become blurred and ambiguous. Some overlap with each other or say different things when read in opposite directions. Many contain inserted sequences of DNA that look as if they have no function (although some of the supposed junk does a useful job, whereas other sections cause disease, should they wake up and shift position). Plenty of questions remain. How important is the part – often a small part – of each gene that codes for proteins compared with the on and off switches, the accelerators and brakes, and the rest of the control machinery? As is the case for many a blockbuster's plot, the genome works, but only just.

The incoherence of the inherited message is exactly what would be expected on Darwin's idea of "descent with modification". Like the eye, the ear and the elbow, the genome shows no element of design, but is instead filled with compromise, contingency and decay. Evolution can now be defined as "genetics plus time". DNA has become the key to how life began, how it changed, and how, from its tangled and primeval roots, when biological promiscuity ruled and information flowed freely among different forms of life, it produced today's vast diversity of species; each a republic of genes, with little or no exchange with its neighbours.

Evolution unites genetics into a single science. The extraordinary progress in technology and (even more important) in cataloguing the information produced makes it possible to infer from a DNA sequence what its protein looks like and what it may do. A new piece of DNA can at once be compared with every one of the millions of others now available to infer from one creature what a gene might do in quite a different one. That approach is powerful indeed. One of the main variants in control of human skin colour was picked up by finding a light-coloured mutation in a fish and it is now routine to study mammalian ageing using the genes responsible in nematode worms. Genetics is the first science to have accelerated by going into reverse. No longer is there a painstaking trudge from the phenotype to the gene. The fact of evolution means that, more and more, we can use familiar organisms such as fruit flies and humans to understand the biology of creatures as distinct as sea anemones and monkeys.

Evolution needs differences, for without them there could be no change with time (and, for that matter, no science of genetics either). Our understanding of its raw material, of mutation, has been transformed in the past few decades. It emerges as an intrinsic property of the genome itself, rather than the result of a series of unfortunate accidents imposed from outside. What once seemed a strict separation of inherited errors from those that occur in body cells has largely broken down, with many cancers emerging from inherited changes in genes, which then undergo further mutation within body cells.

As this book so clearly and comprehensively shows – almost everything has changed. The science of genetics has got far more complicated than it was only a decade ago. Almost without realising, it has gone back a century and more, when the central question was not about the mechanism of inheritance (which seemed simple in much the same way as it appeared straightforward to my undergraduate teachers who, although not quite as wrong as the Victorians, were certainly far from right). To Darwin and his fellows the evolution of form from fertilized egg to adult was a much more interesting issue. Why did the eggs of an elephant or an eel, so similar in their first stages, grow into such spectacularly different creatures? Why, for that matter, were (as Darwin noted) white cats with blue eyes nearly always deaf? And why were hybrids sterile? Many of those questions are addressed in these pages.

The technology expounded here would have amazed Mendel and might even have startled Watson and Crick. Its intricacies and its achievements are well laid out in *The Britannica Guide*. Inevitably, genetics has become big business, with applications in fields as different as medical diagnosis, drug production, law enforcement, and chicken breeding. A science that once turned on ingenuity is now ruled as much by cash as by ideas. Fortunately, the amount needed is declining. It cost billions of dollars to sequence the first human genome. The Archon prize, still unclaimed, offers ten million dollars to the first biologist to read off a hundred human gene sequences in ten days or less. No doubt, someone will soon succeed. Such is progress in DNA technology that, if it continues (and that is not guaranteed) a human genome will be read off in a few hours for a few hundred dollars and a complete sequence will become part of every patient's medical records.

Quite what that might do for health is less clear. Already, a physician can estimate a patient's probable life expectancy within minutes of their entering the surgery. How old are they; and how fat? Male or female? Smoker, drinker, or neither? Income and where he or she lives, history of disease, and – a nod at technology – blood pressure: put all those figures together and, for most people, genetic analysis will add almost nothing to the diagnosis.

Even so, doctors will, no doubt, recommend a quick DNA test. For some people they may be crucial but for most of us they can help. As all geneticists (but few patients) know, inherited variants can alter an individual's susceptibility to environmental challenges. If everyone smoked, lung cancer would be a genetic disease: the combination of inborn vulnerability with the many carcinogens found in a cigarette leads to disaster. Smokers tend not to take much notice, but other drugs – those used against cancer included – also vary greatly in efficacy from one individual to another. As a result, drug treatments made personal by referring to a patient's genes have become part of medicine.

Whatever its promises to medicine, to the study of the past, or to the improvement of food production, genetics has always been the overstated science. It is a source of unreasonable fears and equally unwarranted hopes. Most of the fears are overstated or even plain silly (although nobody working in the field can ignore the crimes committed in the name of eugenics) but some of the hopes are more disturbing. I have myself been approached on more than one occasion by teenagers who tell me that they have cystic fibrosis but that they are not concerned as they will soon be cured by gene therapy – a tale told by teachers who have fallen for the publicity machine built by those anxious to profit from science. Gene therapy may some day revolutionize medicine (and in a few cases has already had some success), genetically modified crops may feed the world, and DNA may explain society, but none of that has yet quite happened.

The Britannica Guide to Genetics is a welcome antidote to such ballyhoo. It gives a clear, unbiased and impressively complete introduction to the science of inheritance. Turn the page and read on.

PART I

BUILDING LIFE

1

MOLECULES OF LIFE

Introduction

Genetics is the study of heredity in general and of genes in particular. Genes are integral to the explanation of hereditary observations. Discoveries into the nature of genes have shown that genes are important determinants of all aspects of an organism's make-up. For this reason, most areas of biological research have a genetic component, and the study of genetics has a position of central importance in biology. Genetic research has also demonstrated that virtually all organisms on this planet have similar genetic systems, with genes that are built on the same chemical principle and that function according to similar mechanisms. Although species differ in the sets of genes they contain, many similar genes are found across a wide range of species. For example, a large proportion of genes in baker's yeast are also present in humans. This similarity in genetic make-up between organisms that have such disparate phenotypes can be explained by the evolutionary relatedness of virtually all life forms on Earth. This genetic unity has radically reshaped the understanding of the relationship between humans and all other organisms. Genetics has also had a profound impact on human affairs. Throughout history humans have created or improved many different medicines, foods, and textiles by subjecting plants, animals, and microbes to the ancient techniques of selective breeding and to the modern methods of recombinant DNA technology. In recent years medical researchers have begun to discover the role that genes play in disease. The significance of genetics only promises to become greater as the structure and function of more and more human genes are characterized.

DNA: Agent of Heredity

Structure and Composition of DNA

The remarkable properties of the nucleic acids, the substances that serve as the carriers of genetic information, have captured the attention of many investigators. The groundwork of genetics was laid by pioneer biochemists who found that nucleic acids are long chain-like molecules, the backbones of which consist of repeated sequences of phosphate and sugar linkages – ribose sugar in ribonucleic acid (RNA) and deoxyribose sugar in deoxyribonucleic acid (DNA). Attached to the sugar links in the backbone are two kinds of nitrogen-containing compounds, or bases: purines and pyrimidines. The purines are adenine (A) and guanine (G) in both DNA and RNA; the pyrimidines are cytosine (C) and thymine (T) in DNA, and cytosine (C) and uracil (U) in RNA. A single purine or pyrimidine is attached to each sugar, and the entire phosphate–sugar–base subunit is called a nucleotide. The nucleic acids extracted from different species of animals and plants have different proportions of the four nucleotides. Some are relatively richer in adenine and thymine, while others have more guanine and cytosine. However, in the early 1950s the Austrian-born American biochemist Erwin Chargaff found that the amount of A is always equal to T, and the amount of G is always equal to C.

With the general acceptance of DNA as the chemical basis of heredity in the early 1950s, many scientists turned their attention to determining how the nitrogenous bases fit together to make up a thread-like molecule. The structure of DNA was determined by American geneticist James Watson and British biophysicist Francis Crick in 1953. Watson and Crick based their model largely on the research of British physicists Rosalind Franklin and Maurice Wilkins, who analysed X-ray diffraction patterns to show that DNA is a double helix. The findings of Chargaff suggested to Watson and Crick that A was somehow paired with T, and G with C.

Using this information, Watson and Crick came up with their now-famous model showing DNA as a double helix composed of two intertwined chains of nucleotides, in which the As of one chain are linked to the Ts of the other, and the Gs in one chain are linked to the Cs of the other (Figure 1.1). The structure resembles a ladder that has been twisted into a spiral shape: the sides of the ladder are composed of sugar and phosphate groups, and the rungs are made up of the paired nitrogenous bases. When Watson and Crick made a wire model of the structure, it became clear that the only way the model could conform to the requirements of the molecular dimensions of DNA was if A always paired with T and G with C; in fact, the A–T and G–C pairs showed a satisfying lock-and-key fit. Although most of the bonds in DNA are strong covalent bonds, the A–T and G–C bonds are weak hydrogen bonds. However, multiple hydrogen bonds along the centre of the molecule confer enough stability to hold the two strands together.

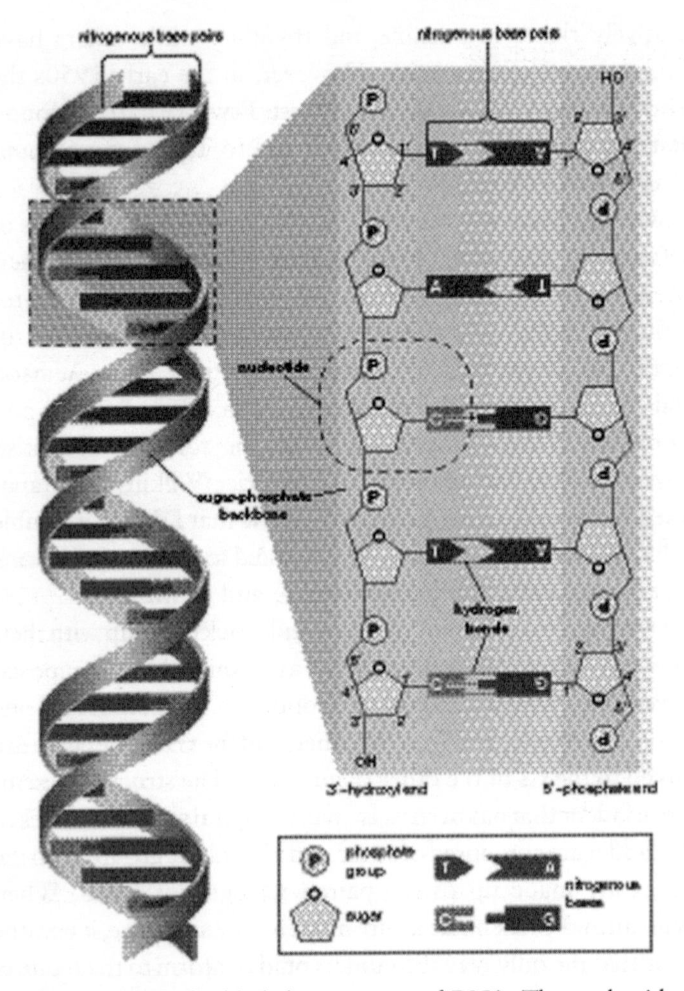

Figure 1.1 The double helix structure of DNA. The nucleotides A, T, G, and C are arranged in opposite orientation, with A pairing with T, and G with C. Watson and Crick noted that the structure fulfilled the features of a hereditary molecule: (1) the double helix could separate, and the two strands could act as templates for replication of identical strands; and (2) the nucleotide sequence could represent coded information.

DNA Replication

The Watson–Crick model of the structure of DNA suggested at least three different ways that DNA might self-replicate. The experiments of American geneticists Matthew Meselson and Franklin Stahl on the bacterium *Escherichia coli* in 1958 suggested that DNA replicates semi-conservatively. Meselson and Stahl grew bacterial cells in the presence of ^{15}N, a heavy atom species (isotope) of nitrogen, so that the DNA of the cells contained ^{15}N. These cells were then transferred to a medium containing the normal isotope of nitrogen, ^{14}N, and allowed to go through cell division. The researchers were able to demonstrate that, in the DNA molecules of the daughter cells, one strand contained only ^{15}N, and the other strand contained ^{14}N. This is precisely what is expected by the semi-conservative mode of replication, in which the original DNA molecules should separate into two template strands containing ^{15}N, and the newly aligned nucleotides should all contain ^{14}N.

The hooking together of free nucleotides in the newly synthesized strand takes place one nucleotide at a time. An incoming free nucleotide pairs with the complementary nucleotide on the template strand, and then one end of the incoming nucleotide is covalently joined to a nucleotide already in place. The process is then repeated. The result is a nucleotide chain, referred to chemically as a nucleotide polymer or a polynucleotide. Of course the polymer is not a random polymer; its nucleotide sequence has been directed by the nucleotide sequence of the template strand. It is this templating process that enables hereditary information to be replicated accurately and passed down through the generations. In a very real way, human DNA has been replicated in a direct line of descent from the first vertebrates that evolved hundreds of millions of years ago.

DNA replication starts at a site on the DNA called the origin of replication. In higher organisms, replication begins at multiple origins of replication and moves along the DNA in both directions outwards from each origin.

In order for DNA to replicate, however, the two strands of the double helix first must be unwound from each other. A class of enzymes called DNA topoisomerases removes helical twists by cutting a DNA strand and then resealing the cut. Enzymes called helicases then separate the two strands of the double helix, exposing two template surfaces for the alignment of free nucleotides. Beginning at the origin of replication, a complex enzyme called DNA polymerase moves along the DNA molecule, pairing nucleotides on each template strand with free complementary nucleotides.

Expression of the Genetic Code

DNA represents a type of information that is vital to the shape and form of an organism. It contains instructions in a coded sequence of nucleotides, and this sequence interacts with the environment to produce form – the living organism with all of its complex structures and functions. The form of an organism is largely determined by protein. A large proportion of the observable parts of an organism are protein; for example, hair, muscle, and skin are made up largely of protein. Other chemical compounds that make up the human body, such as carbohydrates, fats, and more complex chemicals, are either synthesized by enzymes or are deposited at specific times and in specific tissues under the influence of proteins. For example, the black-brown skin pigment melanin is synthesized by enzymes and deposited in special skin cells called melanocytes.

Genes exert their effect mainly by determining the structure and function of the many thousands of different proteins, which in turn determine the characteristics of an organism. Generally, it is true to say that each protein is coded for by one gene, bearing in mind that the production of some proteins requires the cooperation of several genes.

Proteins are polymeric molecules; that is, they are made up of chains of monomeric elements, as is DNA. In proteins, the monomers are amino acids. Organisms generally contain 20 different types of amino acids, and the distinguishing factors that make one protein different from another are its length and specific amino acid sequence, which are determined by the number and sequence of nucleotide pairs in DNA. In other words, the protein polymer has a parallel structure to the DNA polymer, and this is how genetic information flows from DNA into protein.

However, this is not a single-step process. First, the nucleotide sequence of DNA is copied into the nucleotide sequence of single-stranded RNA in a process called transcription. Transcription of any one gene takes place at the chromosomal location of that gene. Whereas the unit of replication is a whole chromosome, the transcriptional unit is a relatively short segment of the chromosome, the gene. The active transcription of a gene depends on the need for the activity of that particular gene in a specific tissue or at a given time.

The nucleotide sequence in RNA faithfully mirrors that of the DNA from which it was transcribed. The U in RNA has exactly the same hydrogen-bonding properties as T in DNA, so there are no changes at the information level. For most RNA molecules, the nucleotide sequence is converted into an amino acid sequence, a process called translation. In prokaryotes, translation begins during the transcription process, before the full RNA transcript is made. In eukaryotes, transcription finishes, and the RNA molecule passes from the nucleus into the cytoplasm, where translation takes place.

The genome of a type of virus called a retrovirus (of which the human immunodeficiency virus, or HIV, is an example) is composed of RNA instead of DNA. In a retrovirus, RNA is reverse transcribed into DNA, which can then integrate into the chromosomal DNA of the host cell that the retrovirus infects. The synthesis of DNA is catalysed by the enzyme reverse transcriptase. The existence of reverse transcriptase shows that genetic information is capable of flowing from RNA to DNA in exceptional cases. Since it is believed that life arose in an RNA world, it is likely that the evolution of reverse transcriptase was an important step in the transition to the present DNA world.

Transcription

A gene is a functional region of a chromosome that is capable of making a transcript in response to appropriate regulatory signals. Therefore, a gene must not only be composed of the DNA sequence that is actually transcribed, but it must also include an adjacent regulatory, or control, region that is necessary for the transcript to be made in the correct developmental context.

The polymerization of ribonucleotides during transcription is catalysed by the enzyme RNA polymerase. As with DNA replication, the two DNA strands must separate to expose the template. However, transcription differs from replication in that for any gene, only one of the DNA strands is actually used as a template. Polymerization is continuous as the RNA polymerase moves along the transcribed region. The RNA strand is extruded from the transcription complex like a tail, which grows longer as the transcription process advances.

Eventually, a full-length transcript of RNA is produced, and this detaches from the DNA. The process is repeated, and multiple RNA transcripts are produced from one gene.

Prokaryotes possess only one type of RNA polymerase, but in eukaryotes there are several different types. RNA polymerase I synthesizes ribosomal RNA (rRNA), and RNA polymerase III synthesizes transfer RNA (tRNA) and other small RNAs. The types of RNA transcribed by these two polymerases are never translated into protein. RNA polymerase II transcribes the major type of genes, those genes that code for proteins. Transcription of these genes is considered in detail below.

Transcription of protein-coding genes results in a type of RNA called messenger RNA (mRNA), so named because it carries a genetic message from the gene on a nuclear chromosome into the cytoplasm, where it is acted upon by the protein-synthesizing apparatus. The transcription machinery contains many items in addition to the RNA polymerase. The successful binding of the RNA polymerase to the DNA "upstream" of the transcribed sequence depends upon the cooperation of many additional proteinaceous transcription factors. The region of the gene upstream from the region to be transcribed contains specific DNA sequences that are essential for the binding of transcription factors and a region called the promoter, to which the RNA polymerase binds. These sequences must be a specific distance from the transcriptional start site for successful operation. Various short base sequences in this regulatory region physically bind specific transcription factors by virtue of a lock-and-key fit between the DNA and the protein. As might be expected, a protein binds with the centre of the DNA molecule, which contains the sequence specificity, and not with the outside of the molecule, which is merely a uniform repetition of sugar and phosphate groups.

In eukaryotes, a key segment is the TATA box, a TATA sequence approximately 30 nucleotides upstream from the transcription start site. RNA polymerase II and numerous other transcription factors assemble in a precise sequence around the TATA box, binding to each other and to the DNA. Together, RNA polymerase and the transcription factors constitute the transcription complex.

The RNA polymerase is directed by the transcription complex to begin transcription at the proper site. It then moves along the template, synthesizing mRNA as it goes. At some position past the coding region, the transcription process stops. Bacteria (prokaryotes) have well-characterized specific termination sequences; however, in eukaryotes, termination signals are less well understood, and the transcription process stops at variable positions past the end of the coding sequence. A short nucleotide sequence downstream from the coding region acts as a signal for the RNA to be cut at that position. Subsequently, approximately 200 adenine nucleotides are added to this end of the newly formed RNA molecule as what is called a poly(A) tail, which is characteristic of all eukaryotic RNA. A modified guanine nucleotide, called a cap, is added at the other end of the mRNA. Nucleotide sequences that cannot be transcribed into proteins (these non-coding sections are called introns) are excised from the RNA at this stage in a process called intron splicing. The adjacent coding regions (known as exons) are then linked together. The resulting tailed, capped, and intron-free molecule is now mature mRNA.

Genetic code

As described above, hereditary information is contained in the nucleotide sequence of DNA in a kind of code. The coded information is copied faithfully into RNA and translated into chains of amino acids. Amino acid chains are folded into helices, zigzags, and other shapes and are sometimes associated with other amino acid chains. The specific amounts of amino acids in a protein and their sequence determine the protein's unique properties. For example, muscle protein and hair protein contain the same 20 amino acids, but the sequences of these amino acids in the two proteins are quite different.

If the nucleotide sequence of mRNA is thought of as a written message, it can be said that this message is read by the translation apparatus in "words" of three nucleotides, starting at one end of the mRNA and proceeding along the length of the molecule. These three-letter words are called codons. Each codon stands for a specific amino acid, so if the message in mRNA is 900 nucleotides long, which corresponds to 300 codons, it will be translated into a chain of 300 amino acids.

Each of the three letters in a codon can be filled by any one of the four nucleotides; therefore, there are 4^3, or 64, possible codons. Each one of these 64 words in the codon dictionary has meaning. Most codons code for one of the 20 possible amino acids. Two amino acids, methionine and tryptophan, are each coded for by one codon only (AUG and UGG, respectively). The other 18 amino acids are coded for by 2 to 6 codons; for example, either of the codons UUU or UUC will cause the insertion of the amino acid phenylalanine into the growing amino acid chain. Three codons – UAG, UGA, and UAA – represent translation-termination signals and are called the stop codons. The first amino acid in an amino acid chain is methionine, encoded by an AUG codon.

One of the surprising findings about the genetic codon

dictionary is that, with a few exceptions, it is the same in all organisms. (One exception is mitochondrial DNA, which exhibits several differences from the standard genetic code and also between organisms.) The uniformity of the genetic code has been interpreted as an indication of the evolutionary relatedness of all organisms. For the purpose of genetic research, codon uniformity is convenient because any type of DNA can be translated in any organism.

Translation

The process of translation requires the interaction not only of large numbers of proteinaceous translational factors but also of specific membranes and organelles of the cell. In both prokaryotes and eukaryotes, translation takes place on cytoplasmic organelles called ribosomes. Ribosomes are aggregations of many different types of proteins and ribosomal RNA (rRNA). They can be thought of as cellular anvils on which the links of an amino acid chain are forged. A ribosome is a generic protein-making machine that can be recycled and used to synthesize many different types of proteins. A ribosome attaches to one end of an mRNA molecule, begins translation at the start codon AUG, and translates the message one codon at a time until a stop codon is reached. Any one mRNA is translated many times by several ribosomes along its length, each one at a different stage of translation. In eukaryotes, ribosomes that produce proteins to be used in the same cell are not associated with membranes. However, proteins that must be exported to another location in the organism are synthesized on ribosomes located on the outside of flattened membranous chambers called the endoplasmic reticulum. A completed amino acid chain is extruded into the inner cavity of the endoplasmic reticulum. Subsequently, the proteins are transported via small vesicles to another organelle called the Golgi apparatus, which in turn buds off more vesicles that eventually fuse with the cell membrane. The protein is then released from the cell (Figure 1.2).

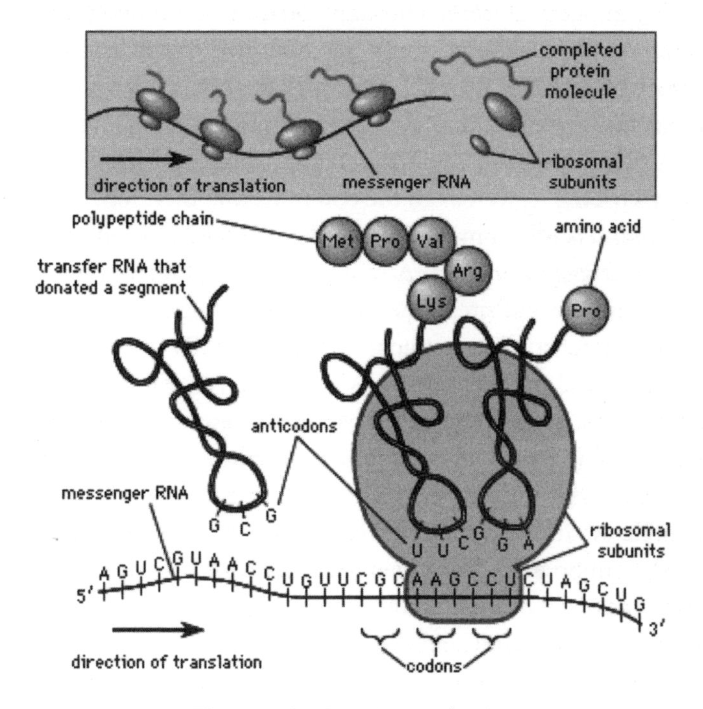

Figure 1.2 Protein synthesis.

Another crucial component of the translational process is transfer RNA (tRNA). The function of any one tRNA molecule is to bind to a designated amino acid and carry it to a ribosome, where the amino acid is added to the growing amino acid chain. Each amino acid has its own set of tRNA molecules that will bind only to that specific amino acid. A tRNA molecule is a single nucleotide chain with several helical regions and a loop containing three unpaired nucleotides, called an anticodon. The anticodon of any one tRNA fits perfectly into the mRNA codon that codes for the amino acid attached to that tRNA; for example, the mRNA codon UUU, which codes for the amino acid phenylalanine, will be bound by the anticodon AAA. Thus, any mRNA codon that happens to be on the ribosome at any one time will solicit the binding only of the tRNA with the appropriate anticodon, which will align the correct amino acid for addition to the chain.

A tRNA molecule and its attached amino acid must bind to the ribosome as well as to the codon during this elongation of the amino acid chain. A ribosome has two tRNA binding sites; at the first site, one tRNA attaches to the amino acid chain, and at the second site, another tRNA carrying the next amino acid is attached. After attachment, the first tRNA departs and recycles, and the second tRNA is now left holding the amino acid chain. At this time the ribosome moves to the next codon, and the whole process is successively repeated along the length of the mRNA until a stop codon is reached, at which time the completed amino acid chain is released from the ribosome.

The amino acid chain then spontaneously folds to generate the three-dimensional shape necessary for its function as a protein. Each amino acid has its own special shape and pattern of electrical charges on its surface, and ultimately these are what determine the overall shape of the protein. The protein's shape is stabilized mainly by weak bonds that form between different parts of the chain. In some proteins, strong covalent bridges are formed between different sites in the chain. If the protein is composed of two or more amino acid chains, these also associate spontaneously and take on their most stable three-dimensional shape. For an enzyme, its shape determines the ability to bind to its specific substrate. For a structural protein, the amino acid sequence determines whether it will be a filament, a sheet, a globule, or another shape.

Gene Mutation

Given the complexity of DNA and the vast number of cell divisions that take place within the lifetime of a multicellular organism, copying errors are likely to occur. If unrepaired, such errors will change the sequence of the DNA bases and alter the genetic code.

Mutation is the random process whereby genes change from one allelic form to another. Scientists who study mutation use the most common genotype found in natural populations, called the wild type, as the standard against which to compare a mutant allele. Mutation can occur in two directions; mutation from wild type to mutant is called a forward mutation, and mutation from mutant to wild type is called a back mutation or reversion.

Mechanisms of mutation

Mutations arise from changes to the DNA of a gene. These changes can be quite small, affecting only one nucleotide pair, or they can be relatively large, affecting hundreds or thousands of nucleotides. Mutations in which one base is changed are called point mutations – for example, substitution of the nucleotide pair AT by GC, CG, or TA.

Base substitutions can have different consequences at the protein level. Some base substitutions are "silent", meaning that they result in a new codon that codes for the same amino acid as the wild type codon at that position, or a codon that codes for a different amino acid that happens to have the same properties as those in the wild type. Substitutions that result in a functionally different amino acid are called "missense" mutations; these can lead to alteration or loss of protein function. A more severe type of base substitution, called a "nonsense" mutation, results in a stop codon in a position where there was not one before, which causes the premature termination of protein synthesis and, more than likely, a complete loss of function in the finished protein.

Another type of point mutation that can lead to drastic loss of function is a frameshift mutation, the addition or deletion of one or more DNA bases. In a protein-coding gene, the sequence of codons starting with AUG and ending with a termination codon is called the reading frame, and it is "read" in groups of three nucleotides at a time. If a nucleotide pair is added to or subtracted from this sequence, the reading frame from that point will be shifted by one nucleotide pair, and all of the codons downstream will be altered. The result will be a protein whose first section (before the mutational site) is that of the wild type amino acid sequence, followed by a tail of functionally meaningless amino acids. Large deletions of many codons will not only remove amino acids from a protein but may also result in a frameshift mutation if the number of nucleotides deleted is not a multiple of three. Likewise, an insertion of a block of nucleotides will add amino acids to a protein and perhaps also have a frameshift effect.

A number of human diseases are caused by the expansion of a trinucleotide pair repeat. For example, fragile-X syndrome, the most common type of inherited intellectual disability in humans, is caused by the repetition of up to 1,000 copies of a CGG repeat in a gene on the X chromosome.

The impact of a mutation depends upon the type of cell involved. In a haploid cell, which has only one set of chromosomes, any mutant allele will most likely be expressed in the phenotype of that cell. In a diploid cell, which has both maternal and paternal chromosomes, a dominant mutation will be expressed over the wild type allele, but a recessive mutation will remain masked by the wild type. If recessive mutations occur in both members of one gene pair in the same cell, the mutant phenotype will be expressed. Mutations in germinal cells (i.e. reproductive cells) may be passed on to successive generations. However, mutations in somatic (body) cells will exert their effect only on that individual and will not be passed on to progeny.

The impact of an expressed somatic mutation depends upon which gene has been mutated. In most cases, the somatic cell with the mutation will die, an event that is generally of little consequence in a multicellular organism. However, mutations in a special class of genes called proto-oncogenes can cause uncontrolled division of that cell, resulting in a group of cells that constitutes a cancerous tumour.

Mutations can affect gene function in several different ways. First, the structure and function of the protein coded by that gene can be affected. For example, enzymes are particularly susceptible to mutations that affect the amino acid sequence at their active site (i.e. the region that allows the enzyme to bind with its specific substrate). This may lead to enzyme inactivity; a protein is made, but it has no enzymatic function. Second, some nonsense or frameshift mutations can lead to the complete absence of a protein. Third, changes to the promoter region of the gene can result in gene malfunction by interfering with transcription. In this situation, protein production is either inhibited or it occurs at an inappropriate time because of alterations somewhere in the regulatory region. Fourth, mutations within introns that affect the specific nucleotide sequences that direct intron splicing may result in an mRNA that still contains an intron. When translated, this extra RNA will almost certainly be meaningless at the protein level, and its extra length will lead to a functionless protein. Any mutation that results in a lack of function for a particular gene is called a "null" mutation. Less severe mutations are called "leaky" mutations because some normal function still "leaks through" into the phenotype.

Most mutations occur spontaneously and have no known cause. The synthesis of DNA is a cooperative venture of many different interacting cellular components, and occasionally mistakes occur that result in mutations. Like many chemical structures, the bases of DNA are able to exist in several conformations called isomers. Occasionally the conformation may change spontaneously to a form that has different base-pairing properties. If this change results in a functionally different amino acid, then a missense mutation may result. Another spontaneous event that can lead to mutation is depurination, the complete loss of a purine base (adenine or guanine) at some location in the DNA. The resulting gap can be filled by any base during subsequent replications.

Researchers have demonstrated that ionizing radiation, some chemicals, and certain viruses are capable of acting as mutagens – agents that can increase the rate at which mutations occur. Some mutagens have been implicated as a cause of cancer. For example, ultraviolet radiation from the sun is known to cause skin cancer, and substances in cigarette smoke are a primary cause of lung cancer.

Repair of mutation

A variety of mechanisms exist for repairing copying errors caused by DNA damage. One of the best-studied systems is the repair mechanism for damage caused by ultraviolet radiation. Ultraviolet radiation joins adjacent Ts, creating T dimers, which, if not repaired, may cause mutations. Special repair enzymes either cut the bond between the Ts or excise the bonded dimer and replace it with two single Ts. If both of these repair methods fail, a third method allows the DNA replication process to bypass the dimer; however, it is this bypass system that causes most mutations because bases are then inserted at random opposite the T dimer. Xeroderma pigmentosum, a severe hereditary disease of humans, is caused by a mutation in a gene coding for one of the T dimer repair enzymes. Individuals with this disease are highly susceptible to skin cancer.

Reverse mutation from the aberrant state of a gene back to its normal, or wild type, state can result in a number of possible molecular changes at the protein level. True reversion is the reversal of the original nucleotide change. However, phenotypic reversion can result from changes that restore a different amino acid with properties identical to the original. Second-site changes within a protein can also restore normal function. For example, an amino acid change at a site different from that altered by the original mutation can sometimes interact with the amino acid at the first mutant site to restore a normal protein shape. Also, second-site mutations at other genes can act as suppressors, restoring wild type function. For example, mutations in the anticodon region of a tRNA gene can result in a tRNA that sometimes inserts an amino acid at an erroneous stop codon; if the original mutation is caused by a stop codon, which arrests translation at that point, then a tRNA anticodon change can insert an amino acid and allow translation to continue normally to the end of the mRNA. Alternatively, some mutations at separate genes open up a new biochemical pathway that circumvents the block of function caused by the original mutation.

Regulation of Gene Expression

Not all genes in a cell are active in protein production at any given time. Gene action can be switched on or off in response to the cell's stage of development and external environment. In multicellular organisms, different kinds of cells express different parts of the genome. In other words, a skin cell and a muscle cell contain exactly the same genes, but the differences in structure and function of these cells result from the selective expression and repression of certain genes.

In prokaryotes and eukaryotes, most gene-control systems are positive, meaning that a gene will not be transcribed unless it is activated by a regulatory protein. However, some bacterial genes show negative control. In this case the gene is transcribed continuously unless it is switched off by a regulatory protein. An example of negative control in prokaryotes involves three adjacent genes used in the metabolism of the sugar lactose by the bacterium *E. coli*. The part of the chromosome containing the genes concerned is divided into two regions, one that includes the structural genes (i.e. those genes that together code for protein structure) and another that is a regulatory region. This overall unit is called an operon. If lactose is not present in a cell, transcription of the genes that code for the lactose-processing enzymes – beta-galactosidase, permease, and transacetylase – is turned off. This is achieved by a protein called the lac repressor, which is produced by the repressor gene and binds to a region of the operon called the operator. Such binding prevents RNA polymerase, which initially binds at the adjacent promoter, from moving into the coding region. If lactose enters the cell, it binds to the lac repressor and induces a change of shape in the repressor so that it can no longer bind to the DNA at the operon. Consequently, the RNA polymerase is able to travel from the promoter down the three adjacent protein-coding regions, making one continuous transcript. This three-gene transcript is subsequently translated into three separate proteins.

Although the operon model has proved a useful model of gene regulation in bacteria, different regulatory mechanisms are employed in eukaryotes. First, there are no operons in eukaryotes, and each gene is regulated independently. Furthermore, the series of events associated with gene expression in higher organisms is much more complex than in prokaryotes and involves multiple levels of regulation.

In order for a gene to produce a functional protein, a complex series of steps must occur. Some type of signal must initiate the transcription of the appropriate region along the DNA, and, finally, an active protein must be made and sent to the appropriate location to perform its specific task. Regulation can be exerted at many different places along this pathway. The fundamental level of control is the rate of transcription. Transcription itself is also a complex process with many different components, and each one is a potential point of control. Regulatory proteins called activators or enhancers are needed for the transcription of genes at a specific time or in a certain cell. Thus, control is positive (not negative as in the lac operon) in that these proteins are necessary for the promotion of transcription. Activators bind to specific regions of the DNA in the upstream regulatory region, some very distant from the binding of the initiation complex.

For most genes intron splicing is a routine step in the production of mRNA, but in some genes there are alternative ways to splice the primary transcript. These result in different mRNAs, which in turn result in different proteins (see Transcription, above).

Some genes are controlled at the translational and post-translational levels. One type of translational control is the storage of uncapped mRNA to meet future demands for protein synthesis. In other cases, control is exerted through the stability or instability of mRNA. The rate of translation of some mRNAs can also be regulated.

Other types of control take place after translation. Certain proteins (e.g. insulin) are synthesized in an inactive form and must be chemically modified to become active; this is known as post-translational modification. Still other proteins are targeted to specific locations inside the cell (e.g. mitochondria) by means of highly specific amino acid sequences at their ends, called leader sequences; when the protein reaches its correct site, the leader segment is cut off, and the protein begins to function. Post-translational control is also exerted through mRNA and protein degradation.

Repetitive DNA

One major difference between the genomes of prokaryotes and eukaryotes is that most eukaryotes contain repetitive DNA, with the repeats either clustered or spread out between the unique genes. There are several categories of repetitive DNA: (1) single copy DNA, which contains the structural genes (protein-coding sequences); (2) families of DNA, in which one gene somehow copies itself, and the repeats are located in small clusters (tandem repeats) or spread throughout the genome (dispersed repeats); and (3) satellite DNA, which contains short nucleotide sequences repeated as many as thousands of times. Such repeats are often found clustered in tandem near the centromeres of the chromosomes during cell division. Microsatellite DNA is composed of tandem repeats of two nucleotide pairs that are dispersed throughout the genome. Minisatellite DNA, sometimes called variable number tandem repeats (VNTRs), is composed of blocks of longer repeats also dispersed throughout the genome. There is no known function for satellite DNA, nor is it known how the repeats are created. There is a special class of relatively large DNA elements called transposons, which can make replicas of themselves that "jump" to different locations in the genome; most transposons eventually become inactive and no longer move, but, nevertheless, their presence contributes to repetitive DNA.

Extranuclear DNA

It was once thought that all of the genetic information in a cell was confined to the DNA in the chromosomes of the cell nucleus. Today, it is known that small circular chromosomes, called extranuclear, or cytoplasmic, DNA, are located in two types of organelles found in the cytoplasm of the cell. These organelles are the mitochondria in animal and plant cells and the chloroplasts in plant cells. Chloroplast DNA (cpDNA) contains genes that are involved with aspects of photosynthesis and with components of the special protein-synthesizing apparatus that is active within the organelle. Mitochondrial DNA (mtDNA) contains some of the genes that participate in the conversion of the energy of chemical bonds into the energy currency of the cell – a chemical called adenosine triphosphate (ATP) – as well as genes for mitochondrial protein synthesis.

The cells of several groups of organisms contain small extra DNA molecules called plasmids. Bacterial plasmids are circular DNA molecules; some carry genes for resistance to various agents in the environment that would be toxic to the bacteria (e.g. antibiotics). Many fungi and some plants possess plasmids in their mitochondria; most of these are linear DNA molecules carrying genes that seem to be relevant only to the propagation of the plasmid and not the host cell.

Polymerase Chain Reaction

The polymerase chain reaction (PCR) technique is used to make numerous copies of a specific segment of DNA quickly and accurately. This technique enables investigators to obtain the large quantities of DNA that are required for various experiments and procedures in molecular biology, forensic analysis, evolutionary biology, and medical diagnostics.

PCR was developed in 1983 by Kary B. Mullis, an American biochemist who won the Nobel Prize for Chemistry in 1993 for his invention. Before the development of PCR, the methods used to amplify, or generate copies of, recombinant DNA fragments were time-consuming and labour-intensive. In contrast, a machine designed to carry out PCR reactions can complete many rounds of replication, producing billions of copies of a DNA fragment, in only a few hours.

The PCR technique is based on the natural processes a cell uses to replicate a new DNA strand. Only a few biological ingredients are needed for PCR. The integral component is the template DNA – i.e. the DNA that contains the region to be copied, such as a gene. As little as one DNA molecule can serve as a template. The only information needed for this fragment to be replicated is the sequence of two short regions of nucleotides (the subunits of DNA) at either end of the region of interest. These two short template sequences must be known, so that two primers – short stretches of nucleotides that correspond to the template sequences – can be synthesized. The primers bind, or anneal, to the template at their complementary sites and serve as the starting point for copying. DNA synthesis at one primer is directed toward the other, resulting in replication of the desired intervening sequence. Also needed are free nucleotides used to build the new DNA strands and a DNA polymerase, an enzyme that does the building by sequentially adding on free nucleotides according to the instructions of the template.

PCR is a three-step process that is carried out in repeated cycles. The initial step is the denaturation, or separation, of the two strands of the DNA molecule. This is accomplished by heating the starting material to temperatures of about 95°C (203°F). Each strand is a template on which a new strand is built. In the second step the temperature is reduced to about

55°C (131°F) so that the primers can anneal to the template. In the third step the temperature is raised to about 72°C (162°F), and the DNA polymerase begins adding nucleotides on to the ends of the annealed primers. At the end of the cycle, which lasts about five minutes, the temperature is raised and the process begins again. The number of copies doubles after each cycle. Usually 25–30 cycles produce a sufficient amount of DNA.

In the original PCR procedure, one problem was that the DNA polymerase had to be replenished after every cycle because it was not stable at the high temperatures needed for denaturation. This problem was solved in 1987 with the discovery of a heat-stable DNA polymerase called *Taq,* an enzyme isolated from the thermophilic bacterium *Thermus aquaticus,* which inhabits hot springs. *Taq* polymerase also led to the invention of the PCR machine.

Because DNA from a wide range of sources can be amplified, the PCR technique has been applied to many fields. It is used to diagnose genetic disease and to detect low levels of viral infection. In forensic medicine it is used to analyse minute traces of blood and other tissues in order to identify the donor by his or her genetic "fingerprint". The technique has also been used to amplify DNA fragments found in preserved tissues, such as those of a 40,000-year-old frozen woolly mammoth or of a 7,500-year-old human found in a peat bog.

The Human Genome

Approximately three billion base pairs of DNA make up the entire set of chromosomes of the human organism. The human genome includes the coding regions of DNA, which encode all the genes (about 25,000) of the human organism, as well as the non-coding regions of DNA, which do not encode any genes. By 2003 the DNA sequence of the entire human genome was known.

The human genome, like the genomes of all other living animals, is a collection of long polymers of DNA. These polymers are maintained in duplicate copy in the form of chromosomes in every human cell and encode in their sequence of constituent bases (G, A, T, and C) the details of the molecular and physical characteristics that form the corresponding organism (see Structure and Composition of DNA, above). The sequence of these polymers, their organization and structure, and the chemical modifications they contain not only provide the machinery needed to express the information held within the genome but also provide the genome with the capability to replicate, repair, package, and otherwise maintain itself.

In addition, the genome is essential for the survival of the human organism; without it no cell or tissue could live for more than a short period of time. For example, red blood cells (erythrocytes), which live for only about 120 days, and skin cells, which on average live for only about 17 days, must be renewed to maintain the viability of the human body, and it is within the genome that the fundamental information for the renewal of these cells, and many other types of cells, is found.

The human genome is not uniform. Excepting identical (monozygous) twins, no two humans share exactly the same genomic sequence. Further, the human genome is not static. Subtle and sometimes not so subtle changes arise with startling frequency. Some of these changes are neutral or even advantageous; these are passed from parent to child and eventually become commonplace in the population. Other changes may be detrimental, resulting in reduced survival or decreased fertility of those individuals who harbour them; these changes tend to be rare in the population. The genome of modern humans, therefore, is a record of the trials and successes of the generations that have come before. Reflected in the variation of the modern genome is the range of diversity that underlies what are typical traits of the human species. There is also evidence in the human genome of the continuing burden of detrimental variations that sometimes lead to disease.

Knowledge of the human genome provides an understanding of the origin of the human species, the relationships between subpopulations of humans, and the health tendencies or disease risks of individual humans. Indeed, in the past 20 years knowledge of the sequence and structure of the human genome has revolutionized many fields of study, including medicine, anthropology, and forensics. With technological advances that enable inexpensive and expanded access to genomic information, the amount of and the potential applications for the information that is extracted from the human genome is extraordinary.

Role of the Human Genome in Research

Since the 1980s there has been an explosion in genetic and genomic research. The combination of the discovery of the polymerase chain reaction, improvements in DNA sequencing technologies, advances in bioinformatics (mathematical biological analysis), and increased availability of faster, cheaper computing power has given scientists the ability to discern and interpret vast amounts of genetic information from tiny samples of biological material. Further, methodologies such as fluorescence in situ hybridization (FISH) and comparative genomic hybridization (CGH) have enabled the detection of the organization and copy number of specific sequences in a given genome.

Understanding the origin of the human genome is of particular interest to many researchers since the genome is indicative of the evolution of humans. The public availability of full or almost full genomic sequence databases for humans and a multitude of other species has allowed researchers to compare and contrast genomic information between individuals, populations, and species. From the similarities and differences observed, it is possible to track the origins of the human genome and to see evidence of how the human species has expanded and migrated to occupy the planet.

Origins of the Human Genome

Comparisons of specific DNA sequences between humans and their closest living relative, the chimpanzee, reveal 99 per cent identity, although the homology drops to 96 per cent if insertions and deletions in the organization of those sequences are taken into account. This degree of sequence variation between humans and chimpanzees is only about 10-fold greater than that seen between two unrelated humans. From other comparisons of the human genome with the genomes of other species, it is clear that the genome of modern humans shares common ancestry with the genomes of all other animals on the planet and that the modern human genome arose between 150,000 and 300,000 years ago.

Comparing the DNA sequences of groups of individuals from different continents allows scientists to define the relationships and even the ages of these different populations. By combining these genetic data with archaeological and linguistic information, anthropologists have been able to discern the origins of *Homo sapiens* in Africa and to track the timing and location of the waves of human migration out of Africa that led to the eventual spread of humans to other continents. For example, genetic evidence indicates that the first humans migrated out of Africa approximately 60,000 years ago, settling in southern Europe, the Middle East, southern Asia, and Australia. From there, subsequent and sequential migrations brought humans to northern Eurasia and across what was then a land bridge to North America and finally to South America.

Social Impacts of Human Genome Research

Databases have been compiled that list and summarize specific DNA variations that are common in certain human populations but not in others. Because the underlying DNA sequences are passed from parent to child in a stable manner, these genetic variations provide a tool for distinguishing the members of one population from those of the other. Public genetic ancestry projects, in which small samples of DNA can be submitted and analysed, have allowed individuals to trace the continental or even subcontinental origins of their most ancient ancestors.

The role of genetics in defining traits and health risks for individuals has been recognized for generations. Long before DNA or genomes were understood, it was clear that many traits tended to run in families and that family history was one of the strongest predictors of health or disease. Knowledge of the human genome has advanced that realization, enabling studies that have identified the genes and even specific sequence variations that contribute to a multitude of traits and disease risks. With this information in hand, health-care professionals are able to practise predictive medicine, which translates in the best of scenarios to preventative medicine. Indeed, pre-symptomatic genetic diagnoses have enabled countless people to live longer and healthier lives. For example, mutations responsible for familial cancers of the breast and colon have been identified, enabling pre-symptomatic testing of individuals in at-risk families. Individuals who carry the mutant gene or genes are counselled to seek heightened surveillance. In this way, if and when cancer appears, these individuals can be diagnosed early, when the cancers are most effectively treated.

Human Genome Project

The Human Genome Project, which operated from 1990 to 2003, was an international collaboration that successfully determined, stored, and rendered publicly available the sequences of almost all the genetic content of the chromosomes of the human organism, otherwise known as the human genome. It provided researchers with basic information about the sequences of the three billion chemical base pairs that make up human genomic DNA. The Human Genome Project was further intended to improve the technologies needed to interpret and analyse genomic sequences, to identify all the approximately 25,000 genes encoded in human DNA, and to address the ethical, legal, and social implications that might arise from defining the entire human genomic sequence.

Timeline

Even before the Human Genome Project, the base sequences of numerous human genes had been determined through contributions made by many individual scientists. However, the vast majority of the human genome remained unexplored, and researchers, having recognized the necessity and value of having at hand the basic information of the human genomic sequence, were beginning to search for ways to uncover this information more quickly. Because the Human Genome Project required billions of dollars that would inevitably be taken away from traditional biomedical research, many scientists, politicians, and ethicists became involved in vigorous debates over the merits, risks, and relative costs of sequencing the entire human genome in one concerted undertaking. Despite the controversy, the Human Genome Project was initiated in 1990 under the leadership of American geneticist Francis Collins, with support from the US Department of Energy and the National Institutes of Health (NIH). The effort was soon joined by scientists from around the world. Moreover, a series of technical advances in the sequencing process itself and in the computer hardware and software used to track and analyse the resulting data enabled rapid progress of the project.

Technological advance, however, was only one of the forces driving the pace of discovery of the Human Genome Project. In 1998 a private-sector enterprise, Celera Genomics, headed by American biochemist and former NIH scientist J. Craig Venter, began to compete with and potentially undermine the publicly funded Human Genome Project. At the heart of the competition was the prospect of gaining control over potential patents on the genome sequence, which was considered a pharmaceutical treasure trove. Although the legal and financial reasons remain unclear, the rivalry between Celera and the NIH ended when they joined forces, thus speeding completion of the rough draft sequence of the human genome. The completion of the rough draft was announced in June 2000 by Collins and Venter. For the next three years, the rough draft sequence was refined, extended, and further analysed, and in April 2003, coinciding with the 50th anniversary of the publication that described the double-helical structure of DNA, written by British biophysicist Francis Crick and American geneticist and biophysicist James D. Watson, the Human Genome Project was declared complete.

2

UNDERSTANDING DNA AND GENES

Areas of Study

Classical Genetics

Classical genetics, which remains the foundation for all other areas in genetics, is concerned primarily with the method by which genetic traits – classified as dominant (always expressed), recessive (subordinate to a dominant trait), intermediate (partially expressed), or polygenic (due to multiple genes) – are transmitted in plants and animals. These traits may be sex-linked (resulting from the action of a gene on the sex, or X, chromosome) or autosomal (resulting from the action of a gene on a chromosome other than a sex chromosome). Classical genetics began with Mendel's study of inheritance in garden peas and continues with studies of inheritance in many different plants and animals. Today, a prime reason for performing classical genetics is for gene discovery – the finding and assembling of a set of genes that affects a biological property of interest.

Microbial Genetics

Microorganisms were generally ignored by the early geneticists because they are small in size and were thought to lack variable traits and the sexual reproduction necessary for a mixing of genes from different organisms. After it was discovered that microorganisms have many different physical and physiological characteristics that are amenable to study, they became objects of great interest to geneticists because of their small size and the fact that they reproduce much more rapidly than larger organisms. Bacteria became important model organisms in genetic analysis, and many discoveries of general interest in genetics arose from their study. Bacterial genetics is the centre of cloning technology.

Viral genetics is another key part of microbial genetics. The genetics of viruses that attack bacteria were the first to be elucidated. Since then, studies and findings of viral genetics have been applied to viruses pathogenic on plants and animals, including humans. Viruses are also used as vectors (agents that carry and introduce modified genetic material into an organism) in DNA technology (see Chapter 9, DNA Cloning).

Population Genetics

The study of genes in populations of animals, plants, and microbes provides information on past migrations, evolutionary relationships and extents of mixing among different varieties and species, and methods of adaptation to the environment. Statistical methods are used to analyse gene distributions and chromosomal variations in populations.

Population genetics is based on the mathematics of the frequencies of alleles (different forms of the same gene) and of genetic types in populations. For example, the Hardy–

Weinberg formula predicts the frequency of individuals with a particular set of genes in a randomly mating population. Selection, mutation, and random changes can be incorporated into such mathematical models to explain and predict the course of evolutionary change at the population level. These methods can be used on alleles of known effect, such as the recessive allele for albinism, or on DNA segments of any type of known or unknown function.

Human population geneticists have traced the origins and migration and invasion routes of modern humans, *Homo sapiens*. DNA comparisons between the present peoples on the planet have pointed to an African origin of *Homo sapiens*. Tracing specific forms of genes has allowed geneticists to deduce probable migration routes out of Africa to the areas colonized today. Similar studies show to what degree present populations have been mixed by recent patterns of travel (see Chapter 4, Human Evolution).

Cytogenetics

Cytogenetics, the microscopic study of chromosomes, blends the skills of cytologists, who study the structure and activities of cells, with those of geneticists, who study genes. Cytologists discovered chromosomes and the way in which they duplicate and separate during cell division at about the same time that geneticists began to understand the behaviour of genes at the cellular level. The close correlation between the two disciplines led to their combination.

Plant cytogenetics early became an important subdivision of cytogenetics because, as a general rule, plant chromosomes are larger than those of animals. Animal cytogenetics became important after the development of the so-called squash technique, in which entire cells are pressed flat on a piece of glass and observed through a microscope; the human chromosomes were numbered using this technique.

Today there are multiple ways to attach molecular labels to specific genes and chromosomes, as well as to specific RNAs and proteins, that make these molecules easily discernible from other components of cells, thereby greatly facilitating cytogenetics research.

Molecular Genetics

Molecular genetics is the study of the molecular structure of DNA, its cellular activities (including its replication), and its influence in determining the overall make-up of an organism. Molecular genetics relies heavily on genetic engineering (recombinant DNA technology – see Chapter 9), which can be used to modify organisms by adding foreign DNA, thereby forming transgenic organisms. Since the early 1980s, these techniques have been used extensively in basic biological research and are also fundamental to the biotechnology industry, which is devoted to the manufacture of agricultural and medical products. Transgenesis forms the basis of gene therapy, the attempt to cure genetic disease by the addition of normally functioning genes from exogenous sources.

Genomics

The development of the technology to sequence the DNA of whole genomes (the complete genetic complement of an organism) on a routine basis has given rise to the discipline of genomics, which dominates genetics research today. Genomics is the study of the structure, function, and evolutionary comparison of whole genomes. Genomics has made it possible to study gene function at a broader level, revealing sets of genes that interact to impinge on some biological property of interest to the researcher. Bioinformatics is the computer-based discipline that deals with the analysis of such large sets of biological information, especially as it applies to genomic information.

Human Genetics

Some geneticists specialize in the hereditary processes of human genetics. Most of the emphasis is on understanding and treating genetic disease and genetically influenced ill health, areas collectively known as medical genetics. One broad area of activity is laboratory research dealing with the mechanisms of human gene function and malfunction and investigating pharmaceutical and other types of treatments. Since there is a high degree of evolutionary conservation between organisms, research on model organisms such as bacteria, fungi, and vinegar flies (*Drosophila*, sometimes called fruit flies) which are easier to study, often provides important insights into human gene function.

Many single-gene diseases, caused by mutant alleles of a single gene, have been discovered. Two well-characterized single-gene diseases are phenylketonuria (PKU) and Tay–Sachs disease. Other diseases, such as heart disease, schizophrenia, and depression, are thought to have more complex hereditary components that involve a number of different genes. These diseases are the focus of a great deal of research that is being carried out today.

Another broad area of activity is clinical genetics, which centres on advising parents of the likelihood of their children being affected by genetic disease caused by mutant genes and abnormal chromosome structure and number. Such genetic counselling is based on examining individual and family medical records and on diagnostic procedures that can detect unexpressed, abnormal forms of genes. Counselling is carried out either by physicians with a particular interest in this area or by specially trained health-care professionals.

Behaviour Genetics

Another aspect of genetics is the study of the influence of heredity on behaviour. Many aspects of animal behaviour are genetically determined and can therefore be treated as similar to other biological properties. This is the subject material of behaviour genetics, whose goal is to determine which genes control various aspects of behaviour in animals. Human behaviour is difficult to analyse because of the powerful effects of environmental factors, such as culture. Few cases of genetic determination of complex human behaviour are known. Genomics studies provide a useful way to explore the genetic factors involved in complex human traits such as behaviour.

Epigenetics

Epigenetics is the study of the chemical modification of specific genes or gene-associated proteins of an organism. Epigenetic modifications can define how the information in genes is expressed and used by cells. The term epigenetics came into general use in the early 1940s, when British embryologist Conrad Waddington used it to describe the interactions between genes and gene products, which direct development and give rise to an organism's phenotype (observable characteristics). Since then, information revealed by epigenetics studies has revolutionized the fields of genetics and developmental biology. Specifically, researchers have uncovered a range of possible chemical modifications to DNA and to proteins called histones that associate tightly with DNA in the nucleus. These modifications can determine when or even if a given gene is expressed in a cell or organism.

It is clear that at least some epigenetic modifications are heritable, passed from parents to offspring, although they are not inherited by the same mechanism as is typical genetic information. Typical genetic information is encoded in the sequences of nucleotides that make up the DNA; this information is therefore passed from generation to generation as faithfully as the DNA replication process is accurate. Epigenetic information, however, is inherited only if the chemical modifications that constitute it are regenerated on newly synthesized DNA or proteins. Some forms of epigenetic modification are faithfully transmitted; however, others may be "erased" or "reset", depending on a variety of factors.

The principal type of epigenetic modification that is understood is methylation (addition of a methyl group). Methylation can be transient and can change rapidly during the lifespan of a cell or organism, or it can be essentially permanent once set early in the development of the embryo. Other largely permanent chemical modifications also play a role; these include histone acetylation (addition of an acetyl group), ubiquitination (the addition of a ubiquitin protein), and phosphorylation (the addition of a phosphoryl group).

The specific location of a given chemical modification can also be important. For example, certain histone modifications distinguish actively expressed regions of the genome from regions that are not highly expressed. These modifications may correlate with chromosome banding patterns generated by staining procedures common in karyotype analyses. Similarly, specific histone modifications may distinguish actively expressed genes from genes that are poised for expression or genes that are repressed in different kinds of cells.

Epigenetic changes not only influence the expression of genes in plants and animals but also enable the differentiation of pluripotent stem cells (cells having the potential to become any of many different kinds of cells; see Chapter 10, Stem Cells). In other words, epigenetic changes allow cells that all share the same DNA and are ultimately derived from one fertilized egg to become specialized – for example, as liver cells, brain cells, or skin cells.

As the mechanisms of epigenetics have become better understood, researchers have recognized that the epigenome – chemical modification at the level of the genome – also influences a wide range of biomedical conditions. This new perception has opened the door to a deeper understanding of normal and abnormal biological processes and has offered the possibility of novel interventions that might prevent or ameliorate certain diseases.

Epigenetic contributions to disease fall into two classes. One class involves genes that are themselves regulated epigenetically, such as the imprinted (parent-specific) genes associated with Angelman syndrome or Prader–Willi syndrome. Clinical outcomes in cases of these syndromes depend on the degree to which an inherited normal or mutated gene is or is not expressed. The other class involves genes whose products participate in the epigenetic machinery and thereby regulate the expression of other genes. For example, the gene *MECP2* (methyl CpG binding protein 2) encodes a protein that binds to specific methylated regions of DNA and contributes to the silencing of those sequences. Mutations that impair the *MECP2* gene can lead to Rett syndrome, a developmental disorder.

Many tumours and cancers are believed to involve epigenetic changes attributable to environmental factors. These changes include a general decrease in methylation, which is thought to contribute to the increased expression of growth-promoting genes, punctuated by gene-specific increases in methylation that are thought to silence tumour-suppressor genes. Epigenetic signalling attributed to environmental factors has also been associated with some characteristics of ageing by researchers that studied the apparently unequal ageing rates in genetically identical twins.

One of the most promising areas of epigenetic investigation involves stem cells. Researchers have understood for some time that epigenetic mechanisms play a key role in defining the "potentiality" of stem cells. As those mechanisms become clearer, it may become possible to intervene and effectively alter the developmental state and even the tissue type of given cells. The implications of this work for future clinical regenerative intervention for conditions ranging from trauma to neurodegenerative disease are profound.

Methods in Genetics

Experimental Breeding

Genetically diverse lines of organisms can be crossed in such a way as to produce different combinations of alleles in one line. For example, parental lines are crossed, producing an F_1 generation, which is then allowed to undergo random mating to produce offspring that have pure-breeding genotypes (i.e. *AA*, *bb*, *cc*, or *DD*). This type of experimental breeding is the origin of new plant and animal lines, which are an important part of making laboratory stocks for basic research. Transgenic commercial lines produced experimentally are called genetically modified organisms (GMOs). Many of the plants and animals used by humans today (e.g. cows, pigs, chickens, sheep, wheat, corn [maize], potatoes, and rice) have been bred in this way (see Chapter 10, Genetically Modified Organisms).

Cytogenetic Techniques

Cytogenetics focuses on the microscopic examination of genetic components of the cell, including chromosomes, genes, and gene products. Older cytogenetic techniques involve placing cells in paraffin wax, slicing thin sections, and preparing them for microscopic study. The newer and faster squash technique involves squashing entire cells and studying their contents. Dyes that selectively stain various parts of the cell are used; the genes, for example, may be located by selectively staining the DNA of which they are composed. Radioactive and fluorescent tags are valuable in determining the location of various genes and gene products in the cell.

Tissue-culture techniques may be used to grow cells before squashing; white blood cells can be grown from samples of human blood and studied with the squash technique. One major application of cytogenetics in humans is in diagnosing abnormal chromosomal complements such as Down syndrome (caused by an extra copy of chromosome 21) and Klinefelter syndrome (occurring in males with an extra X chromosome). Some diagnosis is prenatal, performed on cell samples from amniotic fluid or the placenta.

Biochemical Techniques

Biochemistry is carried out at the cellular or subcellular level, generally on cell extracts. Biochemical methods are applied to the main chemical compounds of genetics – notably DNA, RNA, and protein. Biochemical techniques are used to determine the activities of genes within cells and to analyse substrates and products of gene-controlled reactions. In one approach, cells are ground up and the substituent chemicals are fractionated for further analysis. Special techniques (e.g. chromatography and electrophoresis are used to separate the components of proteins, DNA, and RNA so that inherited differences in their structures can be revealed. For example, there are nearly 700 different kinds, or variants, of human haemoglobin molecules, many of which were identified using basic separation techniques. Radioactively tagged compounds are valuable in studying the biochemistry of whole cells. For example, thymine is a compound found only in DNA; if radioactive thymine is placed in a tissue-culture medium in which cells are growing, it becomes incorporated into the cell's DNA during cell replication. When cells containing radioactive thymine are analysed, the results show that the DNA molecule splits in half during duplication, and each half synthesizes its missing components (see Chapter 1, DNA Replication).

Chemical tests are used to distinguish certain inherited conditions of humans; for example, urinalysis and blood analysis reveal the presence of inherited abnormalities such as phenylketonuria (PKU), cystinuria, alkaptonuria, gout, and galactosaemia. Genomics has provided a battery of diagnostic tests that can be carried out on an individual's DNA. Some of these tests can be applied to fetuses in utero.

Physiological Techniques

Physiological techniques, directed at exploring functional properties of organisms, are also used in genetic investigations. In microorganisms, most genetic variations involve some important cell function. Some strains of one bacterium (*E. coli*), for example, are able to synthesize the vitamin thiamin from simple compounds; others, which lack an enzyme necessary for this synthesis, cannot survive unless thiamin is already present. The two strains can be distinguished by placing them on a thiamin-free mixture: those that grow have the gene for the enzyme, those that fail to grow do not. The technique also is applied to human cells, since many inherited human abnormalities are caused by a faulty gene that fails to produce a vital enzyme; albinism, which results from an inability to produce the pigment melanin in the skin, hair, or iris of the eyes, is an example of an enzyme deficiency in humans.

Molecular Techniques

Although overlapping with biochemical techniques, molecular genetics techniques are deeply involved with the direct study of DNA. This field has been revolutionized by the invention of recombinant DNA technology (see Chapter 9). The DNA of any gene of interest from a donor organism (such as a human) can be cut out of a chromosome and inserted into a vector to make recombinant DNA, which can then be amplified and manipulated, studied, or used to modify the genomes of other organisms.

A fundamental step in recombinant DNA technology is amplification. This is carried out by inserting the recombinant DNA molecule into a bacterial cell, which replicates and produces many copies of the bacterial genome and the recombinant DNA molecule (constituting a DNA clone). A collection of large numbers of clones of recombinant donor DNA molecules is called a genomic library. Such libraries are the starting point for sequencing entire genomes such as the human genome. Today genomes can be scanned for small molecular variants called single nucleotide polymorphisms, or SNPs ("snips"), which act as chromosomal tags to associated specific regions of DNA that have a property of interest and may be involved in a human disease or disorder.

Immunological Techniques

Many substances (e.g. proteins) are antigenic, meaning that when they are introduced into a vertebrate body, they stimulate the production of specific proteins called antibodies. Various antigens exist in red blood cells, including those that make up the major human blood groups (A, B, AB, O). These and other antigens are genetically determined; the study of these substances constitutes the field of immunogenetics. Blood antigens of humans include inherited variations, and the particular combination of antigens in an individual is almost as unique as fingerprints and has been used in such areas as paternity testing (although this approach has been largely supplanted by DNA-based techniques).

Immunological techniques are used in blood group determinations in blood transfusions, in organ transplants, and in determining rhesus incompatibility in childbirth. Specific antigens of the human leukocyte antigen (HLA) genes are correlated with human diseases and predispositions to disease.

Antibodies also have a genetic basis, and their seemingly endless ability to match any antigen presented is based on special types of DNA shuffling processes between antibody genes. Immunology is also useful in identifying specific recombinant DNA clones that synthesize a specific protein of interest.

Mathematical Techniques

Because much of genetics is based on quantitative data, mathematical techniques are used extensively in genetics.

The laws of probability are applicable to cross-breeding and are used to predict frequencies of specific genetic constitutions in offspring. Geneticists also use statistical methods to determine the significance of deviations from expected results in experimental analyses. In addition, population genetics is based largely on mathematical logic – for example, the Hardy–Weinberg equilibrium and its derivatives.

Bioinformatics uses computer-centred statistical techniques to handle and analyse the vast amounts of information accumulating from genome sequencing projects. The computer program scans the DNA looking for genes, determining their probable function based on other similar genes, and comparing different DNA molecules for evolutionary analysis. Bioinformatics has made possible the discipline of systems biology, treating and analysing the genes and gene products of cells as a complete and integrated system.

Applied Genetics

Medicine

Genetic techniques are used in medicine to diagnose and treat inherited human disorders. Knowledge of a family history of conditions such as cancer or various other disorders may indicate a hereditary tendency to develop these diseases. Cells from embryonic tissues reveal certain genetic abnormalities, including enzyme deficiencies, that may be present in newborn babies, thus permitting early treatment. Many countries require a blood test of newborn babies to determine the presence of an enzyme necessary to convert an amino acid, phenylalanine, into simpler products. Phenylketonuria (PKU), which results from lack of this enzyme, causes permanent brain damage if not treated soon after birth.

Many different types of human genetic diseases can be detected in embryos as young as 12 weeks; the procedure involves removal and testing of a small amount of fluid from around the embryo (called amniocentesis) or of tissue from the placenta (called chorionic villus sampling).

Gene therapy is based on modification of defective genotypes by adding functional genes made through recombinant DNA technology. Bioinformatics is being used to "mine" the human genome for gene products that might be candidates for designer pharmaceutical drugs.

Agriculture and Animal Husbandry

In agriculture and animal husbandry, genetic techniques are applied to improve plants and animals. Breeding analysis and transgenic modification using recombinant DNA techniques are routinely used. Animal breeders use artificial insemination to propagate the genes of prize bulls. Prize cows can transmit their genes to hundreds of offspring by hormone treatment, which stimulates the release of many eggs that are collected, fertilized, and transplanted to foster mothers. Several types of mammals can be cloned, meaning that multiple identical copies can be produced of certain desirable types.

Plant geneticists use special techniques to produce new types of plants, such as hybrid grains (i.e. produced by crossing wheat and rye) and plants resistant to destruction by insect and fungal pests.

Plant breeders use the techniques of budding and grafting to maintain desirable gene combinations originally obtained from cross-breeding. Transgenic plant cells can be made into plants by growing the cells on special hormones. The use of the chemical compound colchicine, which causes chromosomes to double in number, has resulted in many new varieties of fruits, vegetables, and flowers. Many transgenic lines of crop plants are commercially advantageous and are being introduced into the market.

Industry

Various industries employ geneticists; the brewing industry, for example, may use geneticists to improve the strains of yeast that produce alcohol. The pharmaceutical industry has developed strains of moulds, bacteria, and other microorganisms high in antibiotic yield. Penicillin and ciclosporin (also known as cyclosporin) from fungi, and streptomycin and ampicillin from bacteria, are some examples.

Biotechnology, based on recombinant DNA technology, is used extensively in industry. "Designer" lines of transgenic bacteria, animals, or plants capable of manufacturing certain commercial products are made and used routinely. Examples of products include pharmaceutical drugs and industrial chemicals such as citric acid.

PART 2

GENETICS THROUGH TIME

3

HISTORICAL BACKGROUND

History of Genetics

Since the earliest times, humankind has recognized the influence of heredity and has applied its principles to the improvement of cultivated crops and domestic animals. A Babylonian tablet more than 6,000 years old, for example, shows pedigrees of horses and indicates possible inherited characteristics. Other old carvings show cross-pollination of date palm trees. Most of the mechanisms of heredity, however, remained a mystery until the 19th century, when genetics as a systematic science began.

In 1869 Swiss chemist Johann Friedrich Miescher extracted a substance containing nitrogen and phosphorus from cell nuclei. The substance was originally called nuclein, but it is now known as DNA. Evidence that DNA acts as the carrier of the genetic information was first firmly demonstrated by exquisitely simple microbiological studies. In 1928 British bacteriologist Frederick Griffith was studying two strains of the bacterium *Streptococcus pneumoniae*; one strain was lethal to mice (virulent) and the other was harmless (avirulent).

Griffith found that mice inoculated with either the heat-killed virulent bacteria or the living avirulent bacteria remained free of infection, but mice inoculated with a mixture of both became infected and died. It seemed as if some chemical "transforming principle" had transferred from the dead virulent cells into the avirulent cells and changed them. In 1944 American bacteriologist Oswald T. Avery and his co-workers found that the transforming factor was DNA. Avery and his research team obtained mixtures from heat-killed virulent bacteria and inactivated either the proteins, polysaccharides (sugar subunits), lipids, DNA, or RNA and added each type of preparation individually to avirulent cells. The only molecular class whose inactivation prevented transformation to virulence was DNA. Therefore, it seemed that DNA, because it could transform the bacteria from avirulent to virulent, must be the hereditary material.

A similar conclusion was reached from the study of bacteriophages, viruses that attack and kill bacterial cells. From a host cell infected by one bacteriophage, hundreds of bacteriophage progeny are produced. In 1952 American biologists Alfred D. Hershey and Martha Chase prepared two populations of bacteriophage particles. In one population, the outer protein coat of the bacteriophage was labelled with a radioactive isotope; in the other, the DNA was labelled. After allowing both populations to attack bacteria, Hershey and Chase found that only when DNA was labelled did the progeny bacteriophage contain radioactivity. Therefore, they concluded that DNA is injected into the bacterial cell, where it directs the synthesis of numerous complete bacteriophages at the expense of the host. In other words, in bacteriophages DNA is the hereditary material responsible for the fundamental characteristics of the virus. Today, it is accepted that, in all living organisms, with the exception of some viruses, genes are composed of DNA.

Genetics arose out of the identification of genes, the fundamental units responsible for heredity. Genetics may be defined as the study of genes at all levels, including the ways in which they act in the cell and the ways in which they are transmitted from parents to offspring. Modern genetics focuses on the chemical substance that genes are made of, DNA, and the ways in which it affects the chemical reactions that constitute the living processes within the cell.

Gene action depends on interaction with the environment. Green plants, for example, have genes containing the information necessary to synthesize the photosynthetic pigment chlorophyll that gives them their green colour. Chlorophyll is synthesized in an environment containing light because the gene for chlorophyll is expressed only when it interacts with light. If a plant is placed in a dark environment, chlorophyll synthesis stops because the gene is no longer expressed.

Genetics as a scientific discipline stemmed from the work of Gregor Mendel in the middle of the 19th century. Mendel suspected that traits were inherited as discrete units, and, although he knew nothing of the physical or chemical nature of genes at the time, his units became the basis for the development of the modern understanding of heredity. All present research in genetics can be traced back to Mendel's discovery of the laws governing the inheritance of traits. The word *gene*, coined in 1909 by Danish botanist Wilhelm Johannsen, has given genetics its name.

Ancient Theories of Pangenesis and Blood in Heredity

Although scientific evidence for patterns of genetic inheritance did not appear until Mendel's work, history shows that humankind must have been interested in heredity long before the dawn of civilization. Curiosity must first have been based on human family resemblances, such as similarity in body structure, voice, gait, and gestures. Such notions were instrumental in the establishment of family and royal dynasties. Early nomadic tribes were interested in the qualities of the animals that they herded and domesticated and, undoubtedly, bred selectively. The first human settlements that practised farming appear to have selected crop plants with favourable qualities. Despite this interest, the first recorded speculations on heredity date from the time of the ancient Greeks; some aspects of their ideas are still considered relevant today.

Hippocrates (*c.* 460–375 BCE), known as the father of medicine, believed in the inheritance of acquired characteristics. To account for this, he devised the hypothesis known as pangenesis. He postulated that all organs of the body of a parent gave off invisible "seeds", which were like miniaturized building components and were transmitted during sexual intercourse, reassembling themselves in the mother's womb to form a baby.

Aristotle (384–322 BCE) emphasized the importance of blood in heredity. He thought that the blood supplied generative material for building all parts of the adult body, and he reasoned that blood was the basis for passing on this generative power to the next generation. In fact, he believed that the male's semen was purified blood and that a woman's menstrual blood was her equivalent of semen. These male and female contributions united in the womb to produce a baby. The blood contained hereditary essences, but he believed that the baby would develop under the influence of these essences, rather than being built from the essences themselves.

Aristotle's ideas about the role of blood in procreation were probably the origin of the still prevalent notion that somehow the blood is involved in heredity. Today people still speak of certain traits as being "in the blood" and of "blood lines" and "blood ties".

The Greek model, which invoked a teeming multitude of substances, or essences, differed from the Mendelian model. Mendel's idea was that distinct differences between individuals are determined by differences in single yet powerful hereditary factors. These single hereditary factors were identified as genes. Copies of genes are transmitted through sperm and egg and guide the development of the offspring. Genes are also responsible for reproducing the distinct features of both parents that are visible in their children.

Preformation and Natural Selection

In the two millennia between the lives of Aristotle and Mendel, few new ideas were recorded on the nature of heredity. In the 17th and 18th centuries the idea of preformation was introduced. Scientists using the newly developed microscopes imagined that they could see miniature replicas of human beings inside sperm heads. French biologist Jean-Baptiste Lamarck invoked the idea of "the inheritance of acquired characters", not as an explanation for heredity but as a model for evolution. He lived at a time when the fixity of species was taken for granted, yet he maintained that this fixity was only found in a constant environment. He enunciated the law of use and disuse, which states that when certain organs become specially developed as a result of some environmental need, then that state of development is hereditary and can be passed on to progeny. He believed that in this way, over many generations, giraffes could arise from deer-like animals that had to keep stretching their necks to reach high leaves on trees.

British naturalist Alfred Russel Wallace originally postulated the theory of evolution by natural selection. However,

Charles Darwin's observations during his circumnavigation of the globe aboard HMS *Beagle* (1831–6) provided evidence for natural selection and his suggestion that humans and animals shared a common ancestry. Many scientists at the time believed in a hereditary mechanism that was a version of the ancient Greek idea of pangenesis, and Darwin's ideas did not appear to fit with the theory of heredity that sprang from the experiments of Mendel.

The Work of Mendel

Before Gregor Mendel, theories for a hereditary mechanism were based largely on logic and speculation, not on experimentation. In his monastery garden, Mendel carried out a large number of cross-pollination experiments between variants of the garden pea, which he obtained as pure-breeding lines. From the precise mathematical ratios of the characteristics of the resulting progeny, he deduced not only the existence of discrete hereditary units (genes) but also that the units were present in pairs in the pea plant and that the pairs separated during gamete (sex cell) formation. Mendel also analysed pure lines that differed in pairs of characters, such as seed colour (yellow versus green) and seed shape (round versus wrinkled). From his results, he deduced the independent assortment of separate gene pairs at gamete formation.

Mendel's success can be attributed in part to his classic experimental approach. He chose his experimental organism well and performed many controlled experiments to collect data. From his results, he developed brilliant explanatory hypotheses and went on to test these hypotheses experimentally. Mendel's methodology established a prototype for genetics that is still used today for gene discovery and for understanding the genetic properties of inheritance.

How the Gene Idea Became Reality

Mendel's genes were only hypothetical entities, factors that could be inferred to exist in order to explain his results. The 20th century saw tremendous strides in the development of the understanding of the nature of genes and how they function. Mendel's publications lay unmentioned in the research literature until 1900, when the same conclusions were reached by several other investigators. Then there followed hundreds of papers showing Mendelian inheritance in a wide array of plants and animals, including humans. It seemed that Mendel's ideas were of general validity. Many biologists noted that the inheritance of genes closely paralleled the inheritance of chromosomes during nuclear divisions, called meiosis, that occur in the cell divisions just prior to gamete formation.

The discovery of linked genes

It seemed that genes were parts of chromosomes. In 1909 this idea was strengthened through the demonstration of parallel inheritance of certain *Drosophila* (a type of vinegar fly) genes on sex-determining chromosomes by American zoologist and geneticist Thomas Hunt Morgan. Morgan and one of his students, Alfred Henry Sturtevant, showed not only that certain genes seemed to be linked on the same chromosome but that the distance between genes on the same chromosome could be calculated by measuring the frequency at which new chromosomal combinations arose (these were proposed to be caused by chromosomal breakage and reunion, also known as crossing-over). In 1916 another student of Morgan's, Calvin Bridges, used *Drosophila* with an extra chromosome to prove beyond reasonable doubt that the only way to explain the abnormal inheritance of certain genes was if they were part of the extra chromosome. American geneticist Hermann Joseph

Müller showed that new alleles (called mutations) could be produced at high frequencies by treating cells with X-rays, the first demonstration of an environmental mutagenic agent (mutations can also arise spontaneously). In 1931, American botanist Harriet Creighton and American scientist Barbara McClintock demonstrated that new allelic combinations of linked genes were correlated with physically exchanged chromosome parts.

Early molecular genetics

In 1908, British physician Archibald Garrod proposed the important idea that the human disease alkaptonuria, and certain other hereditary diseases, were caused by inborn errors of metabolism, providing for the first time evidence that linked genes with molecular action at the cell level. Molecular genetics did not begin in earnest until 1941 when American geneticist George Beadle and American biochemist Edward Tatum showed that the genes they were studying in the fungus *Neurospora crassa* acted by coding for catalytic proteins called enzymes. Subsequent studies in other organisms extended this idea to show that genes generally code for proteins. Soon afterwards, American bacteriologist Oswald Avery, Canadian American geneticist Colin M. Macleod, and American biologist Maclyn McCarty showed that bacterial genes are made of DNA, a finding that was later extended to all organisms.

DNA and the genetic code

A major landmark was attained in 1953 when American geneticist and biophysicist James D. Watson and British molecular biologist Francis Crick devised a double helix model for DNA structure, supported by the X-ray diffraction data of British biophysicists Rosalind Franklin and Maurice Wilkins. This model showed that DNA was capable of self-replication by separating its complementary strands and using them as templates for the synthesis of new DNA molecules. Each of the intertwined strands of DNA was proposed to be a chain of chemical groups called nucleotides, of which there were known to be four types. Because proteins are strings of amino acids, it was proposed that a specific nucleotide sequence of DNA could contain a code for an amino acid sequence and hence protein structure (see Chapter 1).

In 1955, American molecular biologist Seymour Benzer, extending earlier studies in *Drosophila*, showed that the mutant sites within a gene could be mapped in relation to each other. His linear map indicated that the gene itself is a linear structure.

In 1958 the strand-separation method for DNA replication (called the semiconservative method) was demonstrated experimentally for the first time by American geneticists Matthew Meselson and Franklin W. Stahl.

In 1961, Crick and South African biologist Sydney Brenner showed that the genetic code must be read in triplets of nucleotides, called codons. American geneticist Charles Yanofsky showed that the positions of mutant sites within a gene matched perfectly the positions of altered amino acids in the amino acid sequence of the corresponding protein.

In 1966 the complete genetic code of all 64 possible triplet coding units (codons), and the specific amino acids they code for, was deduced by American biochemists Marshall Nirenberg and Har Gobind Khorana. Subsequent studies in many organisms showed that the double helical structure of DNA, the mode of its replication, and the genetic code are the same in virtually all organisms, including plants, animals, fungi, bacteria, and viruses.

In 1961, French biologist François Jacob and French biochemist Jacques Monod established the prototypical model for gene regulation by showing that bacterial genes can be turned on (initiating transcription into RNA) and off through the binding action of regulatory proteins to a region just upstream of the coding region of the gene.

Recombinant DNA technology and the polymerase chain reaction

Technical advances have played an important role in the advance of genetic understanding. In 1970, American microbiologists Daniel Nathans and Hamilton Othanel Smith discovered a specialized class of enzymes (called restriction enzymes) that cut DNA at specific nucleotide target sequences. That discovery allowed American biochemist Paul Berg in 1972 to make the first artificial recombinant DNA molecule by isolating DNA molecules from different sources, cutting them, and joining them together in a test tube. These advances allowed individual genes to be cloned (amplified to a high copy number) by splicing them into self-replicating DNA molecules, such as plasmids (circular DNA elements separate from the chromosomes; see Chapter 1, Extranuclear DNA) or viruses, and inserting these into living bacterial cells. From these methodologies arose the field of recombinant DNA technology that presently dominates molecular genetics.

In 1977 two different methods were invented for determining the nucleotide sequence of DNA: one by American molecular biologists Allan Maxam and Walter Gilbert and the other by British biochemist Fred Sanger. Such technologies made it possible to examine the structure of genes directly by nucleotide sequencing, resulting in the confirmation of many of the inferences about genes that were originally made indirectly.

In the 1970s, Canadian biochemist Michael Smith revolutionized the art of redesigning genes by devising a method for inducing specifically tailored mutations at defined sites within a gene, creating a technique known as site-directed mutagenesis. In 1983, American biochemist Kary B. Mullis invented the polymerase chain reaction (PCR), a method for rapidly detecting and amplifying a specific DNA sequence without cloning it (see Chapter 1, Polymerase Chain Reaction).

In the last decade of the 20th century, progress in recombinant DNA technology and in the development of automated sequencing machines led to the elucidation of complete DNA sequences of several viruses, bacteria, plants, and animals (see Chapter 1, The Human Genome).

History of Evolutionary Theory

Early Ideas

All human cultures have developed their own explanations for the origin of the world and of human beings and other creatures. Traditional Judaism and Christianity explain the origin of living beings and their adaptations to their environments – wings, gills, hands, flowers – as the handiwork of an omniscient God. The philosophers of ancient Greece had their own creation myths. Anaximander proposed that animals could be transformed from one kind into another, and Empedocles speculated that they were made up of various combinations of pre-existing parts. Closer to modern evolutionary ideas were the proposals of early Church Fathers such as Gregory of Nazianzus and Augustine, both of whom maintained that not all species of plants and animals were created by God; rather, some had developed in historical times from God's creations. Their motivation was not biological but religious – it would have been impossible to hold representatives of all species in a single vessel such as Noah's Ark; hence, some species must have come into existence only after the Flood.

The notion that organisms may change by natural processes was not investigated as a biological subject by Christian theologians of the Middle Ages, but it was, usually incidentally, considered as a possibility by many, including Albertus Magnus and his student Thomas Aquinas. Aquinas concluded, after detailed discussion, that the development of living creatures such as maggots and flies from non-living matter such as decaying meat was not incompatible with Christian faith or philosophy. But he left it to others to determine whether this actually happened.

The idea of progress, particularly the belief in unbounded human progress, was central to the Enlightenment of the 18th century, particularly in France among such philosophers as the marquis de Condorcet and Denis Diderot and such scientists as Georges-Louis Leclerc, comte de Buffon. But belief in progress did not necessarily lead to the development of a theory of evolution. Pierre-Louis Moreau de Maupertuis proposed the spontaneous generation and extinction of organisms as part of his theory of origins, but he advanced no theory of evolution – the transformation of one species into another through knowable, natural causes. Buffon, one of the greatest naturalists of the time, explicitly considered – and rejected – the possible descent of several species from a common ancestor. He postulated that organisms arise from organic molecules by spontaneous generation, so that there could be as many kinds of animals and plants as there are viable combinations of organic molecules.

The British physician Erasmus Darwin, grandfather of Charles Darwin, offered in his *Zoonomia; or, The Laws of Organic Life* (1794–6) some evolutionary speculations, but they were not further developed and had no real influence on subsequent theories. The Swedish botanist Carolus Linnaeus devised the hierarchical system of plant and animal classification that is still in use in a modernized form. Although he insisted on the fixity of species, his classification system eventually contributed much to the acceptance of the concept of common descent.

The great French naturalist Jean-Baptiste de Monet, chevalier de Lamarck, held the enlightened view of his age that living organisms represent a progression, with humans as the highest form. From this idea he proposed, in the early years of the 19th century, the first broad theory of evolution. Organisms evolve through aeons of time from lower to higher forms, a process still going on, always culminating in human beings. As organisms become adapted to their environments through their habits, modifications occur. Use of an organ or structure reinforces it; disuse leads to obliteration. The characteristics acquired by use and disuse, according to this theory, would be inherited. This assumption, later called the inheritance of acquired characteristics (or Lamarckism), was thoroughly disproved in the 20th century. Although his theory did not stand up in the light of later knowledge, Lamarck made important contributions to the gradual acceptance of biological evolution and stimulated countless later studies.

Charles Darwin

The founder of the modern theory of evolution was Charles Darwin. The son and grandson of physicians, he enrolled as a medical student at the University of Edinburgh. After two years, however, he left to study at the University of Cambridge and prepare to become a clergyman. He was not an exceptional student, but he was deeply interested in natural history. On 27 December 1831, a few months after his graduation from Cambridge, he sailed as a naturalist aboard the HMS *Beagle* on a round-the-world trip that lasted until October 1836. During the voyage, Darwin was often able to disembark for extended trips ashore to collect natural specimens.

The discovery of fossil bones from large extinct mammals in Argentina and the observation of numerous species of finches in the Galapagos islands were among the events credited with stimulating Darwin's interest in how species originate. In 1859 he published *On the Origin of Species by Means of Natural Selection*, a treatise establishing the theory of evolution and, most important, the role of natural selection in determining its course. He published many other books as well, notably *The Descent of Man and Selection in Relation to Sex* (1871), which extends the theory of natural selection to human evolution.

Darwin must be seen as a great intellectual revolutionary who inaugurated a new era in the cultural history of humankind, an era that was the second and final stage of the Copernican revolution that had begun in the 16th and 17th centuries under the leadership of men such as Nicolaus Copernicus, Galileo, and Isaac Newton. The Copernican revolution marked the beginnings of modern science. Discoveries in astronomy and physics overturned traditional conceptions of the universe. Earth was no longer seen as the centre of the universe but as a small planet revolving around one of myriad stars; the seasons and the rains that make crops grow, as well as destructive storms and other vagaries of weather, became understood as aspects of natural processes; the revolutions of the planets were explained by simple laws that also accounted for the motion of projectiles on Earth.

The significance of these and other discoveries was that they led to a conception of the universe as a system of matter in motion governed by laws of nature. The workings of the universe no longer needed to be attributed to the ineffable will of a divine Creator; rather, they were brought into the realm of science – an explanation of phenomena through natural laws. Physical phenomena such as tides, eclipses, and positions of the planets could be predicted whenever the causes were adequately known. Darwin accumulated evidence showing that evolution had occurred, that diverse organisms share common ancestors, and that living beings have changed drastically over the course of Earth's history. More important, however, he extended to the living world the idea of nature as a system of matter in motion governed by natural laws.

Before Darwin, the origin of Earth's living things, with their marvellous contrivances for adaptation, had been attributed to the design of an omniscient God. He had created the fish in the waters, the birds in the air, and all sorts of animals and plants on the land. God had endowed these creatures with gills for breathing, wings for flying, and eyes for seeing, and he had coloured birds and flowers so that human beings could enjoy them and recognize God's wisdom. Christian theologians, from Aquinas on, had argued that the presence of design, so evident in living beings, demonstrates the existence of a supreme Creator; the argument from design was Aquinas's "fifth way" for proving the existence of God. In 19th-century England the eight Bridgewater Treatises were commissioned so that eminent scientists and philosophers would expand on the marvels of the natural world and thereby set forth "the Power, wisdom, and goodness of God as manifested in the Creation".

The British theologian William Paley in his *Natural Theology* (1802) used natural history, physiology, and other contemporary knowledge to elaborate the argument from design. If a person should find a watch, even in an uninhabited desert, Paley contended, the harmony of its many parts would force him to conclude that it had been created by a skilled watchmaker; and, Paley went on, how much more intricate and perfect in design is the human eye, with its transparent lens, its retina placed at the precise distance for forming a distinct image, and its large nerve transmitting signals to the brain.

The argument from design seems to be forceful. A ladder is made for climbing, a knife for cutting, and a watch for telling time; their functional design leads to the conclusion that they have been fashioned by a carpenter, a smith, or a watchmaker. Similarly, the obvious functional design of animals and plants seems to denote the work of a Creator. It was Darwin's genius that he provided a natural explanation for the organization and functional design of living beings. Darwin accepted the facts of adaptation – hands are for grasping, eyes for seeing, lungs for breathing. But he showed that the multiplicity of plants and animals, with their exquisite and varied adaptations, could be explained by a process of natural selection, without recourse to a Creator or any designer agent. This achievement would prove to have intellectual and cultural implications more profound and lasting than his multipronged evidence that convinced contemporaries of the fact of evolution.

Darwin's theory of natural selection is summarized in the *Origin of Species* as follows:

> As many more individuals are produced than can possibly survive, there must in every case be a struggle for existence, either one individual with another of the same species, or with the individuals of distinct species, or with the physical conditions of life. . . . Can it, then, be thought improbable, seeing that variations useful to man have undoubtedly occurred, that other variations useful in some way to each being in the great and complex battle of life, should sometimes occur in the course of thousands of generations? If

> such do occur, can we doubt (remembering that many more individuals are born than can possibly survive) that individuals having any advantage, however slight, over others, would have the best chance of surviving and of procreating their kind? On the other hand, we may feel sure that any variation in the least degree injurious would be rigidly destroyed. This preservation of favourable variations and the rejection of injurious variations, I call Natural Selection.

Natural selection was proposed by Darwin primarily to account for the adaptive organization of living beings; it is a process that promotes or maintains adaptation. Evolutionary change through time and evolutionary diversification (multiplication of species) are not directly promoted by natural selection, but they often ensue as by-products of natural selection as it fosters adaptation to different environments.

Modern Conceptions

The Darwinian aftermath

The publication of the *Origin of Species* produced considerable public excitement. Scientists, politicians, clergymen, and notables of all kinds read and discussed the book, defending or deriding Darwin's ideas. The most visible actor in the controversies immediately following publication was the British biologist T.H. Huxley, known as "Darwin's bulldog", who defended the theory of evolution with articulate and sometimes mordant words on public occasions as well as in numerous writings. Evolution by natural selection was indeed a favourite topic in society salons during the 1860s and beyond. But serious scientific controversies also arose, first in Britain and then on the European continent and in the United States.

One occasional participant in the discussion was the British naturalist Alfred Russel Wallace, who had hit upon the idea of natural selection independently and had sent a short manuscript about it to Darwin from the Malay archipelago, where he was collecting specimens and writing. On 1 July 1858, one year before the publication of the *Origin*, a paper jointly authored by Wallace and Darwin was presented, in the absence of both, to the Linnean Society in London – with apparently little notice. Greater credit is duly given to Darwin than to Wallace for the idea of evolution by natural selection; Darwin developed the theory in considerably more detail, provided far more evidence for it, and was primarily responsible for its acceptance. Wallace's views differed from Darwin's in several ways, most importantly in that Wallace did not think natural selection sufficient to account for the origin of human beings, which in his view required direct divine intervention.

A younger British contemporary of Darwin, with considerable influence during the latter part of the 19th and in the early 20th century, was Herbert Spencer. A philosopher rather than a biologist, he became an energetic proponent of evolutionary ideas, popularized a number of slogans, such as "survival of the fittest" (which was taken up by Darwin in later editions of the *Origin*), and engaged in social and metaphysical speculations. His ideas considerably damaged proper understanding and acceptance of the theory of evolution by natural selection. Darwin wrote of Spencer's speculations:

> His deductive manner of treating any subject is wholly opposed to my frame of mind. ... His fundamental generalizations (which have been compared in importance by some persons with Newton's laws!) which I dare say may be very valuable under a philosophical point of view, are of such a nature that they do not seem to me to be of any strictly scientific use.

The most serious difficulty facing Darwin's evolutionary theory was the lack of an adequate theory of inheritance that would account for the preservation through the generations of the variations on which natural selection was supposed to act. Contemporary theories of "blending inheritance" proposed that offspring merely struck an average between the characteristics of their parents. But as Darwin became aware, blending inheritance (including his own theory of pangenesis, in which each organ and tissue of an organism throws off tiny contributions of itself that are collected in the sex organs and determine the configuration of the offspring) could not account for the conservation of variations, because differences between variant offspring would be halved each generation, rapidly reducing the original variation to the average of the pre-existing characteristics.

The missing link in Darwin's argument was provided by Mendelian genetics. About the time the *Origin of Species* was published, the Augustinian monk Gregor Mendel was starting a long series of experiments with peas in the garden of his monastery in Brünn, Austria-Hungary (now Brno, Czech Republic). These experiments and the analysis of their results are by any standard an example of masterly scientific method. Mendel's paper, published in 1866 in the *Proceedings* of the Natural Science Society of Brünn, formulated the fundamental principles of the theory of heredity that is still current. His theory accounts for biological inheritance through particulate factors (now known as genes) inherited one from each parent, which do not mix or blend but segregate in the formation of the sex cells, or gametes.

Mendel's discoveries remained unknown to Darwin, however, and, indeed, they did not become generally known until 1900, when they were simultaneously rediscovered by a number of scientists in Europe and in the United States. In the meantime, Darwinism in the latter part of the 19th century faced an alternative evolutionary theory known as neo-Lamarckism. This hypothesis shared with Lamarck's the importance of use and disuse in the development and obliteration of organs, and it added the notion that the environment acts directly on organic structures, which explained their adaptation to the way of life and environment of the organism. Adherents of this theory discarded natural selection as an explanation for adaptation to the environment.

Prominent among the defenders of natural selection was the German biologist August Weismann, who in the 1880s published his germ plasm theory. He distinguished two substances that make up an organism: the soma, which comprises most body parts and organs, and the germ plasm, which contains the cells that give rise to the gametes and hence to progeny. Early in the development of an egg, the germ plasm becomes segregated from the somatic cells that give rise to the rest of the body. This notion of a radical separation between germ plasm and soma – that is, between the reproductive tissues and all other body tissues – prompted Weismann to assert that inheritance of acquired characteristics was impossible, and it opened the way for his championship of natural selection as the only major process that would account for biological evolution. Weismann's ideas became known after 1896 as neo-Darwinism.

The synthetic theory

The rediscovery in 1900 of Mendel's theory of heredity, by the Dutch botanist and geneticist Hugo de Vries and others, led to an emphasis on the role of heredity in evolution. De Vries proposed a new theory of evolution known as mutationism, which essentially did away with natural selection as a major evolutionary process. According to de Vries (who was joined by other geneticists such as William Bateson in England), two kinds of variation take place in organisms. One is the "ordinary" variability observed among individuals of a species, which is of no lasting consequence in evolution because, according to de Vries, it could not "lead to a transgression of the species border [i.e. to establishment of new species] even under conditions of the most stringent and continued selection." The other consists of the changes brought about by mutations, spontaneous alterations of genes that result in large modifications of the organism and give rise to new species: "The new species thus originates suddenly, it is produced by the existing one without any visible preparation and without transition."

Mutationism was opposed by many naturalists and in particular by the so-called biometricians, led by the British statistician Karl Pearson, who defended Darwinian natural selection as the major cause of evolution through the cumulative effects of small, continuous, individual variations (which the biometricians assumed passed from one generation to the next without being limited by Mendel's laws of inheritance).

The controversy between mutationists (also referred to at the time as Mendelians) and biometricians approached a resolution in the 1920s and 1930s through the theoretical work of geneticists. These scientists used mathematical arguments to show, first, that continuous variation (in such characteristics as body size, number of eggs laid, and the like) could be explained by Mendel's laws and, second, that natural selection acting cumulatively on small variations could yield major evolutionary changes in form and function. Distinguished members of this group of theoretical geneticists were R.A. Fisher and J.B.S. Haldane in Britain and Sewall Wright in the United States. Their work contributed to the downfall of mutationism and, most important, provided a theoretical framework for the integration of genetics into Darwin's theory of natural selection.

Yet their work had a limited impact on contemporary biologists for several reasons: it was formulated in a mathematical language that most biologists could not understand; it was almost exclusively theoretical, with little empirical corroboration; and it was limited in scope, largely omitting many issues, such as speciation (the process by which new species are formed), that were of great importance to evolutionists.

A major breakthrough came in 1937 with the publication of *Genetics and the Origin of Species* by Theodosius Dobzhansky, a Russian-born American naturalist and experimental geneticist. Dobzhansky's book advanced a reasonably comprehensive account of the evolutionary process in genetic terms, laced with experimental evidence supporting the theoretical argument. *Genetics and the Origin of Species* may be considered the most important landmark in the formulation of what came to be known as the synthetic theory of evolution, effectively combining Darwinian natural selection and Mendelian genetics. It had an enormous impact on naturalists and experimental biologists, who rapidly embraced the new understanding of the evolutionary process as one of genetic change in populations. Interest in evolutionary studies was greatly stimulated, and contributions to the theory soon began to follow, extending the synthesis of genetics and natural selection to a variety of biological fields.

The main writers who, together with Dobzhansky, may be considered the architects of the synthetic theory were the German-born American zoologist Ernst Mayr, the British zoologist Julian Huxley, the American palaeontologist George Gaylord Simpson, and the American botanist George Ledyard Stebbins. These researchers contributed to a burst of evolutionary studies in the traditional biological disciplines and in some emerging ones – notably population genetics and, later, evolutionary ecology. By 1950 acceptance of Darwin's theory of evolution by natural selection was universal among biologists, and the synthetic theory had become widely adopted.

Molecular biology and earth sciences

The most important line of investigation after 1950 was the application of molecular biology to evolutionary studies. In 1953 James Watson and Francis Crick deduced the molecular structure of DNA, the hereditary material contained in the chromosomes of every cell's nucleus (see Chapter 1). The genetic information is encoded within the sequence of nucleotides that make up the chain-like DNA molecules. This information determines the sequence of amino acid building blocks of protein molecules, which include, among others, structural proteins such as collagen, respiratory proteins such as haemoglobin, and numerous enzymes responsible for the organism's fundamental life processes. Genetic information contained in the DNA can thus be investigated by examining the sequences of amino acids in the proteins.

In the mid-1960s laboratory techniques such as electrophoresis and selective assay of enzymes became available for the rapid and inexpensive study of differences among enzymes and other proteins. The application of these techniques to evolutionary problems made possible the pursuit of issues that earlier could not be investigated – for example, exploring the extent of genetic variation in natural populations (which sets bounds on their evolutionary potential) and determining the amount of genetic change that occurs during the formation of new species.

Comparisons of the amino acid sequences of corresponding proteins in different species provided quantitatively precise measures of the divergence among species evolved from common ancestors, a considerable improvement over the typically qualitative evaluations obtained by comparative anatomy and other evolutionary sub-disciplines. In 1968 the Japanese geneticist Motoo Kimura proposed the neutrality theory of molecular evolution, which assumes that, at the level of the sequences of nucleotides in DNA and of amino acids in proteins, many changes are adaptively neutral; they have little or no effect on the molecule's function and thus on an organism's fitness within its environment.

If the neutrality theory is correct, there should be a "molecular clock" of evolution; that is, the degree to which amino acid or nucleotide sequences diverge between species should provide a reliable estimate of the time since the species diverged. This would make it possible to reconstruct an evolutionary history that would reveal the order of branching of different lineages, such as those leading to humans, chimpanzees, and orangutans, as well as the time in the past when the lineages split from one another. During the 1970s and 1980s, it gradually became clear that the molecular clock is not exact; nevertheless, into the early 21st century it continued to provide the most reliable evidence for reconstructing evolutionary history. (See Chapter 5, Molecular Clock of Evolution, for a more detailed discussion.)

The laboratory techniques of DNA cloning and sequencing have provided a powerful means of investigating evolution at the molecular level. The fruits of this technology began to accumulate during the 1980s following the development of automated DNA-sequencing machines and the invention of PCR, a simple and inexpensive technique that obtains, in a few hours, billions or trillions of copies of a specific DNA sequence or gene. Major research efforts such as the Human Genome Project further improved the technology for obtaining long DNA sequences rapidly and inexpensively (see Chapter 1).

By the first few years of the 21st century, the full DNA sequence – the full genetic complement, or genome – had been obtained for more than 20 eukaryotes, including humans, the house mouse *Mus musculus*, the rat *Rattus norvegicus*, the vinegar fly *Drosophila melanogaster*, the mosquito *Anopheles gambiae*, the nematode worm *Caenorhabditis elegans*, the malaria parasite *Plasmodium falciparum*, the mustard weed *Arabidopsis thaliana*, and the yeast *Saccharomyces cerevisiae*, as well as for numerous microorganisms (prokaryotes).

In the second half of the 20th century, the earth sciences also experienced a conceptual revolution with considerable consequences for the study of evolution. The theory of plate tectonics, which was formulated in the late 1960s, revealed that the configuration and position of the continents and oceans are dynamic, rather than static, features of the Earth. Oceans grow and shrink, while continents break into fragments or coalesce into larger masses. The continents move across Earth's surface at rates of a few centimetres a year, and over millions of years of geological history this movement profoundly alters the face of the planet, causing major climatic changes along the way. These massive modifications of Earth's past environments are, of necessity, reflected in the evolutionary history of life. Biogeography, the evolutionary study of plant and animal distribution, has been revolutionized by the knowledge, for example, that Africa and South America were part of a single land mass some 200 million years ago and that the Indian subcontinent was not connected with Asia until geologically recent times.

Ecology, the study of the interactions of organisms with their environments, has evolved from descriptive studies – "natural history" – into a vigorous biological discipline with a strong mathematical component, both in the development of theoretical models and in the collection and analysis of quantitative data. Evolutionary ecology is an active field of evolutionary biology; another is evolutionary ethology, the study of the evolution of animal behaviour. Sociobiology, the evolutionary study of social behaviour, is perhaps the most active subfield of ethology. It is also the most controversial, because of its extension to human societies.

Scientific acceptance and extension to other disciplines

The theory of evolution makes statements about three different, though related, issues: (1) the fact of evolution – that is, that organisms are related by common descent; (2) evolutionary history – the details of when lineages split from one another and of the changes that occurred in each lineage; and (3) the mechanisms or processes by which evolutionary change occurs.

The first issue is the most fundamental and the one established with utmost certainty. Darwin gathered much evidence in its support, but evidence has accumulated continuously ever since, derived from all biological disciplines. The evolutionary origin of organisms is today a scientific conclusion established with the kind of certainty attributable to such scientific concepts as the roundness of the Earth, the motions of the planets, and the molecular composition of matter. This degree of certainty beyond reasonable doubt is what is implied when biologists say that evolution is a "fact"; the evolutionary origin of organisms is accepted by virtually every biologist.

But the theory of evolution goes far beyond the general affirmation that organisms evolve. The second and third issues – seeking to ascertain evolutionary relationships between particular organisms and the events of evolutionary history, as well as to explain how and why evolution takes place – are matters of active scientific investigation. Some conclusions are well established. One, for example, is that the chimpanzee and the gorilla are more closely related to humans than is any of those three species to the baboon or other monkeys. Another conclusion is that natural selection, the process postulated by Darwin, explains the configuration of such adaptive features as the human eye and the wings of birds. Many matters are less certain, others are conjectural, and still others – such as the characteristics of the first living things and when they came about – remain unknown.

Since Darwin, the theory of evolution has gradually extended its influence to other biological disciplines, from physiology to ecology and from biochemistry to systematics. All biological knowledge includes the phenomenon of evolution. In the words of Theodosius Dobzhansky, "Nothing in biology makes sense except in the light of evolution."

4

HEREDITY AND EVOLUTION

Introduction

Heredity is the sum of all biological processes by which particular characteristics are transmitted from parents to their offspring. The concept of heredity encompasses two seemingly paradoxical observations about organisms: the constancy of a species from generation to generation and the variation among individuals within a species. Constancy and variation are actually two sides of the same coin, as becomes clear in the study of genetics. Both aspects of heredity can be explained by genes. Every member of a species has a set of genes specific to that species. It is this set of genes that provides the constancy of the species. Among individuals within a species, however, variations can occur in the form each gene takes, providing the genetic basis for the fact that no two individuals (except identical twins) have exactly the same traits.

The set of genes that an offspring inherits from both parents, a combination of the genetic material of each, is called the organism's genotype. The phenotype, on the other hand, is the organism's outward appearance and the developmental outcome of its genes. The phenotype includes an organism's bodily structures, physiological processes, and behaviours. Although the genotype determines the broad limits of the features an organism can develop, the features that actually develop – the phenotype – depend on complex interactions between genes and their environment. The genotype remains constant throughout an organism's lifetime, but its phenotype can change because the organism's internal and external environments change continuously. In conducting genetic studies, it is crucial to discover the degree to which the observable trait is attributable to the pattern of genes in the cells and to what extent it arises from environmental influence.

Basic Features of Heredity

Mendelian Genetics

Discovery and rediscovery of Mendel's laws

Gregor Mendel published his work in the proceedings of the local society of naturalists in Brünn, Austria (now Brno, Czech Republic), in 1866, but none of his contemporaries appreciated its significance. It was not until 1900, 16 years after Mendel's death, that his work was rediscovered independently by botanists Hugo de Vries in Holland, Carl Erich Correns in Germany, and Erich Tschermak von Seysenegg in Austria. Like several investigators before him, Mendel experimented on hybrids of different varieties of a plant; he focused on the common pea plant (*Pisum sativum*). His methods differed in two essential respects from those of his predecessors. First, instead of trying to describe the appearance of whole plants with all their characteristics, Mendel followed the inheritance of single, easily visible and distinguishable traits, such as round versus wrinkled seed, yellow versus green seed, purple versus white flowers, and so on. Second, he made exact counts of the numbers of plants bearing each trait; it was from such quantitative data that he deduced the rules governing inheritance.

Since pea plants usually reproduce by self-pollination of their flowers, the varieties Mendel obtained from seeds were "pure" – descended for several to many generations from plants with similar traits. Mendel crossed them by deliberately transferring the pollen of one variety to the pistils of another; the resulting first-generation hybrids, denoted by the symbol F_1, usually showed the traits of only one parent. For example, the crossing of yellow-seeded plants with green-seeded ones gave yellow seeds, and the crossing of purple-flowered plants with white-flowered ones gave purple-flowered plants. Traits such as the yellow-seed colour and the purple-flower colour Mendel called dominant; the green-seed colour and the white-flower colour he called recessive. It looked as if the yellow and purple characteristics overcame or consumed the green and white ones.

That this was not so became evident when Mendel allowed the F_1 hybrid plants to self-pollinate and produce the second hybrid generation, F_2. Here, both the dominant and the recessive traits reappeared, as pure and uncontaminated as they were in the original parents (generation P). Moreover, these traits appeared in constant proportions: about three-quarters of the plants in the second generation showed the dominant trait and one-quarter showed the recessive, a 3 : 1 ratio.

Mendel concluded that the sex cells, the gametes, of the purple-flowered plants carried some factor that caused the progeny to develop purple flowers, and the gametes of the white-flowered variety had a variant factor that induced the development of white flowers. In 1909 the Danish biologist Wilhelm Johannsen proposed calling these factors genes.

An example of one of Mendel's experiments will illustrate how the genes are transmitted and in what particular ratios. *R* stands for the gene for purple flowers and *r* represents the gene for white flowers (dominant genes are conventionally symbolized by capital letters and recessive genes by lower-case letters). Since each pea plant contains a gene endowment, half of whose set is derived from the mother and half from the father, each plant has two genes for flower colour. If the two genes are alike – for instance, both having come from white-flowered parents (*rr*) – the plant is termed a homozygote. The union of gametes with different genes gives a hybrid plant, termed a heterozygote (*Rr*). Since the gene *R*, for purple, is dominant over *r*, for white, the F_1 generation hybrids will show purple flowers. They are phenotypically purple, but their genotype contains both *R* and *r* genes, and these alternative (allelic or allelomorphic) genes do not blend with or contaminate each other.

Mendel inferred that, when a heterozygote forms its sex cells, the allelic genes segregate and pass to different gametes. This is expressed in Mendel's first law, the law of segregation of unit genes (Figure 4.1). Equal numbers of gametes, ovules, or pollen grains are formed that contain the genes *R* and *r*. If the gametes unite at random, then the F_2 generation should contain about one-quarter white-flowered and three-quarters purple-flowered plants. The white-flowered plants, which must be recessive homozygotes, bear the genotype *rr*. About one-third of the plants exhibiting the dominant trait of purple flowers must be homozygotes, *RR*, and two-thirds heterozygotes, *Rr*. The prediction is tested by obtaining a third generation, F_3, from the purple-flowered plants; though phenotypically all purple-flowered, two-thirds of this F_2 group of plants reveal the presence of the recessive gene allele, *r*, in their genotype by producing about one-quarter white-flowered plants in the F_3 generation.

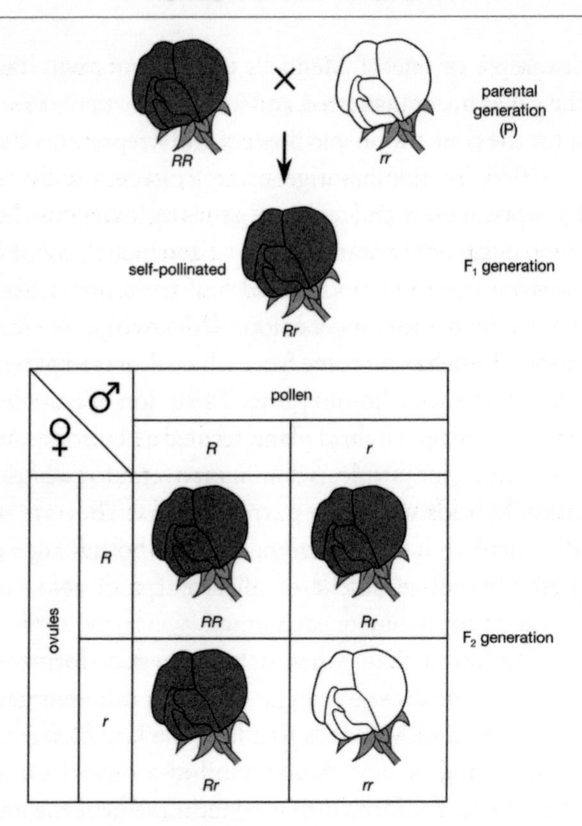

Figure 4.1 Mendel's law of segregation. Cross of a coloured and a white-flowered strain of peas; *R* stands for the gene for coloured flowers and *r* for the gene for white flowers.

Mendel also cross-bred varieties of peas that differed in two or more easily distinguishable traits. When a variety with yellow round seed was crossed to a variety with green wrinkled seed, the F_1 generation hybrids produced yellow round seed. Evidently, yellow (A) and round (B) are dominant traits, and green (a) and wrinkled (b) are recessive. By allowing the F_1 plants (genotype *AaBb*) to self-pollinate, Mendel obtained an F_2 generation of 315 yellow round, 101 yellow wrinkled, 108 green round, and 32 green wrinkled seeds, a ratio of approximately 9 : 3 : 3 : 1. The important point here is that the segregation of the colour (*A–a*) is independent of the segregation of the trait of seed surface (*B–b*). This is expected if the F_1 generation produces equal numbers of four kinds of gametes, carrying the four possible combinations of the parental genes: *AB*, *Ab*, *aB*, and *ab*. Random union of these gametes gives, then, the four phenotypes in a ratio 9 dominant–dominant : 3 recessive–dominant : 3 dominant–recessive : 1 recessive–recessive.

Among these four phenotypic classes there must be nine different genotypes, a supposition that can be tested experimentally by raising a third hybrid generation (F_3). The predicted genotypes are actually found. Another test is by means of a back-cross (or test-cross); the F_1 hybrid (phenotype yellow round seed; genotype *AaBb*) is crossed to a double recessive plant (phenotype green wrinkled seed; genotype *aabb*). If the hybrid gives four kinds of gametes in equal numbers and if all the gametes of the double recessive are alike (*ab*), the predicted progeny of the back-cross are yellow round, yellow wrinkled, green round, and green wrinkled seed in a ratio 1 : 1 : 1 : 1. This prediction is realized in experiments. When the varieties crossed differ in 3 genes, the F_1 hybrid forms 2^3, or 8, kinds of gametes (2^n = kinds of gametes, *n* being the number of genes). The second generation of hybrids, the F_2, has 27 (3^3) genotypically distinct kinds of individuals but only 8 different phenotypes. From these results and others, Mendel derived his second law: the law of recombination, or independent assortment of genes (Figure 4.2).

Universality of Mendel's laws

Although Mendel experimented with varieties of peas, his laws have been shown to apply to the inheritance of many kinds of characters in almost all organisms. In 1902 Mendelian

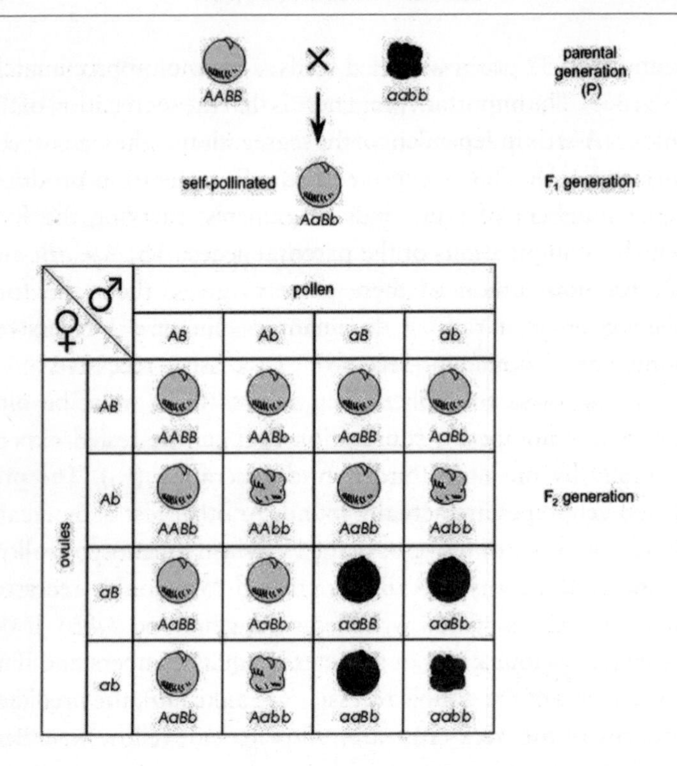

Figure 4.2 Mendel's law of independent assortment. Cross of peas having yellow round seeds with peas having green wrinkled seeds. *A* stands for the gene for yellow and *a* for the gene for green; *B* stands for the gene for a round surface and *b* for the gene for a wrinkled surface.

inheritance was demonstrated in poultry (by British geneticists William Bateson and Reginald Punnett) and in mice. The following year, albinism became the first human trait shown to be a Mendelian recessive, with pigmented skin the corresponding dominant.

In 1902 and 1909, British physician Sir Archibald Garrod published papers on inborn errors of metabolism in humans that provided evidence for the involvement of biochemical genetics. Alkaptonuria, inherited as a recessive disorder, is characterized by excretion in the urine of large amounts of the substance called alkapton, or homogentisic acid, which renders the urine black on exposure to air. Normally the homogentisic acid is converted to acetoacetic acid by the enzyme homogentisic acid oxidase. Garrod advanced the hypothesis that this enzyme is absent or inactive in homozygous carriers of the defective recessive alkaptonuria gene; hence, the homogentisic acid accumulates and is excreted in the urine. Mendelian inheritance of numerous traits in humans has been studied since then.

In analysing Mendelian inheritance, it should be borne in mind that an organism is not an aggregate of independent traits, each determined by one gene. A "trait" is an abstraction, a term of convenience in description. One gene may affect many traits (a condition termed pleiotropy). The white gene in *Drosophila* flies is pleiotropic; it affects the colour of the eyes and of the testicular envelope in the males, the fecundity and the shape of the spermatheca in the females, and the longevity of both sexes. In humans many diseases caused by a single defective gene will have a variety of symptoms, all pleiotropic manifestations of the gene.

Allelic interactions: dominance relationships

The operation of Mendelian inheritance is frequently more complex than in the case of the traits recorded by Mendel. In the first place, clear-cut dominance and recessiveness are by no means always found. When red- and white-flowered varieties of four-o'clock plants (*Mirabilis jalapa*) are crossed, for example, the F_1 hybrids have flowers of intermediate pink or rose colour, a situation that seems more explicable by the blending notion of inheritance than by Mendelian concepts. That the inheritance of flower colour is indeed due to Mendelian mechanisms becomes apparent when the F_1 hybrids are allowed to cross, yielding an F_2 generation of red-, pink-, and white-flowered plants in a ratio of 1 red : 2 pink : 1 white. Obviously the hereditary information for the production of red and white flowers had not been blended away in the F_1 generation, as flowers of these colours were produced in the F_2 generation of hybrids.

The apparent blending in the F_1 generation is explained by the fact that the gene alleles that govern flower colour in *M. jalapa* show an incomplete dominance relationship. Suppose then that a gene allele R_1 is responsible for red flowers and R_2 for white; the homozygotes R_1R_1 and R_2R_2 are red and white respectively, and the heterozygotes R_1R_2 have pink flowers. A similar pattern of lack of dominance is found in Shorthorn cattle. In diverse organisms, dominance ranges from complete (a heterozygote indistinguishable from one of the homozygotes) to incomplete (heterozygotes exactly intermediate) to excessive or overdominance (a heterozygote more extreme than either homozygote).

Another form of dominance is one in which the heterozygote displays the phenotypic characteristics of both alleles. This is called co-dominance; an example is seen in the MN blood group system of human beings. MN blood type is governed by two alleles, *M* and *N*. Individuals who are homozygous for the *M* allele have a surface molecule (called the M antigen) on their red blood cells. Similarly, those homozygous for the *N* allele have the N antigen on the red blood cells. Heterozygotes – those with both alleles – carry both antigens.

Allelic interactions: multiple alleles

The traits discussed so far all have been governed by the interaction of two possible alleles. Many genes, however, are represented by multiple allelic forms within a population. (One individual, of course, can possess only two of these multiple alleles.) Human blood groups – in this case, the well-known ABO system – again provide an example. The gene that governs ABO blood types has three alleles: I^A, I^B, and I^O. I^A and I^B are co-dominant, but I^O is recessive. Because of the multiple alleles and their various dominance relationships, there are four phenotypic ABO blood types: type A (genotypes I^AI^A and I^AI^O), type B (genotypes I^BI^B and I^BI^O), type AB (genotype I^AI^B), and type O (genotype I^OI^O).

Gene interactions

Many individual traits are affected by more than one gene. For example, the coat colour in many mammals is determined by numerous genes interacting to produce the result. The great variety of colour patterns in cats, dogs, and other domesticated animals is the result of different combinations of interacting genes. The gradual unravelling of their modes of inheritance was one of the active fields of research in the early years of genetics.

Two or more genes may produce similar and cumulative effects on the same trait. In humans the skin-colour difference between so-called black people and so-called white people is due to several (probably four or more) interacting pairs of genes, each of which increases or decreases the skin pigmentation by a relatively small amount.

Epistatic genes

Some genes mask the expression of other genes just as a fully dominant allele masks the expression of its recessive counterpart. A gene that masks the phenotypic effect of another gene is called an epistatic gene; the gene it subordinates is the hypostatic gene. The gene for albinism (lack of pigment) in humans is an epistatic gene. It is not part of the interacting skin-colour genes described above; rather, its dominant allele is necessary for the development of any skin pigment, and its recessive homozygous state results in the albino condition regardless of how many other pigment genes may be present. Albinism thus occurs in some individuals among dark- or intermediate-pigmented peoples as well as among light-pigmented peoples.

The presence of epistatic genes explains much of the variability seen in the expression of such dominantly inherited human diseases as Marfan syndrome and neurofibromatosis. Because of the effects of an epistatic gene, some individuals who inherit a dominant, disease-causing gene show only partial symptoms of the disease; some in fact may show no expression of the disease-causing gene, a condition referred to as non-penetrance. The individual in whom such a non-penetrant mutant gene exists will be phenotypically normal but still capable of passing the deleterious gene on to offspring, who may exhibit the full-blown disease.

Examples of epistasis abound in non-human organisms. In mice, as in humans, the gene for albinism has two variants: the allele for non-albino and the allele for albino. The latter allele is unable to synthesize the pigment melanin. Mice, however, have another pair of alleles involved in melanin placement. These are the agouti allele, which produces dark melanization of the hair except for a yellow band at the tip, and the black allele, which produces melanization of the whole hair. If melanin cannot be formed (the situation in the mouse homozygous for the albino gene), neither agouti nor black can be expressed. Hence, homozygosity for the albinism gene is epistatic to the agouti/black alleles and prevents their expression.

Complementation

The phenomenon of complementation is another form of interaction between non-allelic genes. For example, there are mutant genes that in the homozygous state produce profound deafness in humans. One would expect that the children of two persons with such hereditary deafness would be deaf. This is frequently not the case, because the parents' deafness is often caused by different genes. Since the mutant genes are not alleles, the child becomes heterozygous for the two genes and hears normally. In other words, the two mutant genes complement each other in the child. Complementation thus becomes a test for allelism. In the case of congenital deafness cited above, if all the children had been deaf, one could assume that the deafness in each of the parents was owing to mutant genes that were alleles. This would be more likely to occur if the parents were genetically related (consanguineous).

Polygenic inheritance

The greatest difficulties of analysis and interpretation are presented by the inheritance of many quantitative or continuously varying traits. Inheritance of this kind produces variations in degree rather than in kind, in contrast to the inheritance of discontinuous traits resulting from single genes of major effect. The yield of milk in different breeds of cattle; the egg-laying capacity in poultry; and the stature, shape of the head, blood pressure, and intelligence in humans range in continuous series from one extreme to the other and are significantly dependent on environmental conditions. Crosses of two varieties differing in such characters usually give F_1 hybrids intermediate between the parents. At first sight this situation suggests a blending inheritance through "blood" rather than Mendelian inheritance; in fact, it was probably observations of this kind of inheritance that suggested the folk idea of "blood theory".

It has, however, been shown that these characters are polygenic – determined by several or many genes. Each of these characters, taken separately, produces only a slight effect on the phenotype, as small as or smaller than that caused by environmental influences on the same characters. That Mendelian segregation does take place with polygenes, as with the genes having major effects (sometimes called oligogenes), is shown by the variation among F_2 and further-generation hybrids being usually much greater than that in the F_1 generation. By selecting among the segregating progenies the desired variants – for example, individuals with the greatest yield, the best size, or a desirable behaviour – it is possible to produce new breeds or varieties sometimes exceeding the parental forms. Hybridization and selection are consequently potent methods that have been used for improvement of agricultural plants and animals.

Polygenic inheritance also applies to many of the birth defects (congenital malformations) seen in humans. Although expression of the defect itself may be discontinuous (as in club foot, for example), susceptibility to the trait is continuously variable and follows the rules of polygenic inheritance. When a developmental threshold produced by a polygenically inherited susceptibility and a variety of environmental factors is exceeded, the birth defect results.

RNA and non-Mendelian inheritance

In 2006 Minoo Rassoulzadegan from the University of Nice–Sophia Antipolis, France, and her colleagues reported the first indication that a kind of non-Mendelian genetic inheritance originally described in the 1950s for plants also occurred in mammals. The potential implications of the finding were profound, because they concerned the overall understanding of genetic inheritance and how genes are expressed.

Many traits in species that range from plants to humans are inherited in a Mendelian fashion, but by the early 21st century it was clear that most traits in most species follow so-called complex (not strictly Mendelian) patterns of inheritance. Complex traits can result from the combined effects of multiple individual genes, combinations of genetic and environmental influences, or various molecular effects such as DNA instability or histone methylation (a chemical change in a chromosome protein; see Chapter 2, Epigenetics).

The pattern of non-Mendelian inheritance found by Rassoulzadegan and her fellow researchers was called paramutation and involved a case in which a trait but not its corresponding allele was passed from a heterozygous parent to its offspring. The researchers worked with a strain of wild-type (normal) mice and related heterozygous mice that carried an engineered mutant allele of a gene called *Kit*. Each wild-type mouse had a tail that was uniform in colour; the heterozygous mutants had spotted tails. According to Mendel's law of segregation, the expected outcome of a cross (mating) between a normal (homozygous) mouse and a heterozygous mutant mouse would be a litter in which one-half of the pups had spotted tails and the other half did not. Instead, the researchers found spotted tails on all of the pups, including those that carried a pair of normal *Kit* alleles. These mice were paramutated – they showed the trait for the mutant *Kit* allele even though they did not carry it.

To uncover the mechanism of the paramutation, the researchers tested and ruled out a number of logical possibilities, such as DNA or histone methylation. They then explored the levels and structures of *Kit* mRNA in wild-type mice, heterozygous mice, and paramutated mice. The researchers saw diminished levels and degraded forms of *Kit* mRNA in the tissues of both the heterozygous and paramutated animals, and surmised that this effect was transmitted from a heterozygous parent to a paramutated pup through both eggs and sperm, because the effect appeared to be transmitted equally well from females and males.

Further study showed that *Kit* RNA, which was not found in the mature sperm of the wild-type mice, was present in the mature sperm of the heterozygous animals, and it suggested that the RNA in the sperm consisted of small RNA fragments called microRNA, which was known to target corresponding full-length mRNAs for degradation. As a test, the researchers injected a solution of *Kit* microRNAs into otherwise wild-type mouse embryos at the single-cell stage. The pups that developed were paramutated, and even the offspring of the paramutated pups had spotted tails. The fact that microRNAs in early embryos could cause a permanent and heritable change in gene expression meant that this unusual mechanism might account for some fraction of the as-yet poorly understood diversity of traits observed in humans and other animals.

Heredity and Environment

Preformism and epigenesis

A notion that was widespread among pioneer biologists in the 18th century was that the fetus, and hence the adult organism that develops from it, is preformed in the sex cells. Some early microscopists even imagined that they saw a tiny homunculus, a diminutive human figure, encased in the human spermatozoon. The development of the individual from the sex cells appeared deceptively simple: it was merely an increase in the size and growth of what was already present in the sex cells.

The antithesis of the early preformation theory was the theory of epigenesis, which claimed that the sex cells were structureless jelly and contained nothing at all in the way of rudiments of future organisms. The naive early versions of preformation and epigenesis had to be given up when embryologists showed that the embryo develops by a series of complex but orderly and gradual transformations. Darwin's "Provisional Hypothesis of Pangenesis" was distinctly preformistic; Weismann's theory of determinants in the germ plasm, as well as the early ideas about the relations between genes and traits, also tended toward preformism.

Heredity has been defined as a process that results in the progeny's resembling his or her parents. A further qualification of this definition states that what is inherited is a potential that expresses itself only after interacting with and being modified by environmental factors. In short, all phenotypic expressions have both hereditary and environmental components, the amount of each varying for different traits. Thus, a trait that is primarily hereditary (e.g. skin colour in humans) may be modified by environmental influences (e.g. suntanning). And conversely, a trait sensitive to environmental modifications (e.g. weight in humans) is also genetically conditioned.

Organic development is preformistic insofar as a fertilized egg cell contains a genotype that conditions the events that may occur and is epigenetic insofar as a given genotype allows a variety of possible outcomes. These considerations should dispel the reluctance felt by many people to accept the fact that mental as well as physiological and physical traits in humans are genetically conditioned. Genetic conditioning does not mean that heredity is the "dice of destiny". At least in principle, but not invariably in practice, the development of a trait may be manipulated by changes in the environment.

Heritability

Although hereditary diseases and malformations are, unfortunately, by no means uncommon in the aggregate, no one of them occurs very frequently. The characteristics by which one person is distinguished from another – such as facial features, stature, shape of the head, voice, and the colour of skin, eye, and hair – are not usually inherited in a clear-cut Mendelian manner, as are some hereditary malformations and diseases. This is not as strange as it may seem. The kinds of gene changes, or mutations, that produce morphological or physiological effects tend to be drastic enough to be clearly set apart from the more usual phenotypes.

The variations that occur among healthy persons are, as a general rule, caused by polygenes with individually small effects. The same is true of individual differences among members of various animal and plant species. Even brown–blue eye colour in humans, which in many families behaves as if caused by two forms of a single gene (brown being dominant and blue recessive), is often blurred by minor gene modifiers of the pigmentation. Some apparently blue-eyed persons actually carry the gene for the brown eye colour, but several additional modifier genes decrease the amount of brown pigment in the iris. This type of genetic process can influence susceptibility to many diseases (e.g. diabetes) or birth defects (e.g. cleft lip – with or without cleft palate).

The question geneticists must often attempt to answer is how much of the observed diversity between individuals of any species is because of hereditary, or genotypic, variations and how much of it is because of environmental influences. Applied to human beings, this is sometimes referred to as the nature–nurture problem. With animals or plants the problem is evidently more easily soluble than it is with people. Two complementary approaches are possible. First, individual organisms or their progenies are raised in environments as uniform as can be provided, with food, temperature, light, humidity, etc., carefully controlled. The differences that persist between such individuals or progenies probably reflect genotypic differences. Second, individuals with similar or identical genotypes are placed in different environments. The phenotypic differences may then be ascribed to environmental induction. Experiments combining both approaches have been carried out on several species of plants that grow naturally at different altitudes, from sea level to the alpine zone of the Sierra Nevada in California. Young yarrow plants (*Achillea*) were cut into three parts, and the cuttings were replanted in experimental gardens at sea level, at mid-altitude (4,800 feet [1,460 metres]), and at high altitude (10,000 feet [3,050 metres]). It was observed that the plants native at sea level grow best in their native habitat, grow less well at mid-altitudes, and die at high altitudes. On the other hand, the alpine plants survive and develop better at the high-altitude transplant station than at lower altitudes.

With organisms that cannot survive being cut into pieces and placed in controlled environments, a partitioning of the observed variability into genetic and environmental components may be attempted by other methods. Suppose that in a certain population individuals vary in stature, weight, or some other trait. These characters can be measured in many pairs of parents and in their progenies raised under different environmental conditions. If the variation is owing entirely to environment and not at all to heredity, then the expression of the character in the parents and in the offspring will show no correlation (heritability = 0). On the other hand, if the environment is unimportant and the character is uncomplicated by dominance, then the means of this character in the progenies will be the same as the means of the parents; with differences in the expression in females and in males taken into account, the heritability will be 1. In reality, most heritabilities are found to lie between 0 and 1.

It is important to understand clearly the meaning of heritability estimates. They show that, given the range of the environments in which the experimental animals live, one could predict the average body sizes in the progenies of pigs better than one could predict the average numbers of piglets in a litter. The heritability is, however, not an inherent or unchangeable property of each character. If one could make the environments more uniform, the heritabilities would rise, and with more diversified environments they would decrease. Similarly, in populations that are more variable genetically, the heritabilities increase, and in genetically uniform ones, they decrease. In humans the situation is even more complex, because the environments of the parents and of their children are in many ways interdependent. Suppose, for example, that one wishes to study the heritability of stature, weight, or susceptibility to tuberculosis. The stature, weight, and liability to contract tuberculosis depend to some extent on the quality of nutrition and generally on the economic well-being of the family. If no allowance is made for this fact, the heritability estimates arrived at may be spurious; such heritabilities have indeed been claimed for such things as administrative, legal, or military talents and for social eminence in general. It is evident that having socially eminent parents makes it easier for the children to achieve such eminence also; biological heredity may have little or nothing to do with this.

A general conclusion from the evidence available may be stated as follows: diversity in almost any trait – physical, physiological, or behavioral – is due in part to genetic variables and in part to environmental variables. In any array of environments, individuals with more nearly similar genetic endowments are likely to show a greater average resemblance than the carriers of more diverse genetic endowments. It is, however, also true that in different environments the carriers of similar genetic endowments may grow, develop, and behave in different ways.

Physical Basis of Heredity

When Gregor Mendel formulated his laws of heredity, he postulated a particulate nature for the units of inheritance. What exactly these particles were he did not know. Today, scientists understand not only the physical location of hereditary units (i.e. the genes) but their molecular composition as well. The unravelling of the physical basis of heredity makes up one of the most fascinating chapters in the history of biology.

As has been discussed, each individual in a sexually reproducing species inherits two alleles for each gene, one from each parent. Furthermore, when such an individual forms sex cells, each of the resultant gametes receives one member of each allelic pair. The formation of gametes occurs through a process of cell division called meiosis. When gametes unite in fertilization, the double dose of hereditary material is restored, and a new individual is created. This individual, consisting at first of only one cell, grows via mitosis, a process of repeated cell divisions. Mitosis differs from meiosis in that each daughter cell receives a full copy of all the hereditary material found in the parent cell.

It is apparent that the genes must physically reside in cellular structures that meet two criteria. First, these structures must be replicated and passed on to each generation of daughter cells during mitosis. Second, they must be organized into homologous pairs, one member of which is parcelled out to each gamete formed during meiosis.

As early as 1848, biologists had observed that cell nuclei resolve themselves into small rod-like bodies during mitosis. Later, these structures were found to absorb certain dyes and so came to be called chromosomes (coloured bodies). During the early years of the 20th century, cellular studies using ordinary light microscopes clarified the behaviour of chromosomes during mitosis and meiosis, which led to the conclusion that chromosomes are the carriers of genes.

Behaviour of Chromosomes During Cell Division

During mitosis

When the chromosomes condense during cell division, they have already undergone replication. Each chromosome thus consists of two identical replicas, called chromatids, joined at a point called the centromere. During mitosis the sister chromatids separate, one going to each daughter cell. Chromosomes thus meet the first criterion for being the repository of genes: they are replicated, and a full copy is passed to each daughter cell during mitosis.

During meiosis

It was the behaviour of chromosomes during meiosis, however, that provided the strongest evidence for their being the carriers of genes. In 1902 American scientist Walter S. Sutton reported on his observations of the action of chromosomes during sperm formation in grasshoppers. Sutton had observed that, during meiosis, each chromosome (consisting of two chromatids) becomes paired with a physically similar chromosome. These homologous chromosomes separate during meiosis, with one member of each pair going to a different cell. Assuming that one member of each homologous pair was of maternal origin and the other was paternally derived, here was an event that fulfilled the behaviour of genes postulated in Mendel's first law (see Figure 4.1).

It is now known that the number of chromosomes within the nucleus is usually constant in all individuals of a given species – for example, 46 in the human; 40 in the house mouse; 8 in the vinegar fly (*Drosophila melanogaster*); 20 in corn (maize); 24 in the tomato; and 48 in the potato. In sexually reproducing organisms, this number is called the diploid number of chromosomes, as it represents the double dose of chromosomes received from two parents. The nucleus of a gamete, however, contains half this number of chromosomes, or the haploid number. Thus, a human gamete contains 23 chromosomes, while a *Drosophila* gamete contains 4. Meiosis produces the haploid gametes.

The essential features of meiosis are shown in Figure 4.3. For the sake of simplicity, the diploid parent cell is shown to contain a single pair of homologous chromosomes, one member of which is from the father (represented in white in the figure) and the other from the mother (in black). At the leptotene stage the chromosomes appear as long, thin threads. At pachytene they pair, the corresponding portions of the two chromosomes lying side by side. The chromosomes then duplicate and contract into paired chromatids. At this stage the pair of chromosomes is known as a tetrad, as it consists of four chromatids. Also at this stage an extremely important event occurs: portions of the maternal and paternal chromosomes are exchanged. This exchange process, called crossing over, results in chromatids that include both paternal and maternal genes and consequently introduces new genetic combinations. The first meiotic division separates the chromosomal tetrads, with the paternal chromosome (whose chromatids now contain some maternal genes) going to one cell and the maternal chromosome (containing some paternal genes) going to another cell. During the second meiotic division the chromatids separate. The original diploid cell has thus given rise to four haploid gametes (only two of which are shown in the diagram). Not only has a reduction in chromosome number occurred, but the resulting single member of each homologous chromosome pair may be a new combination (through crossing over) of genes present in the original diploid cell.

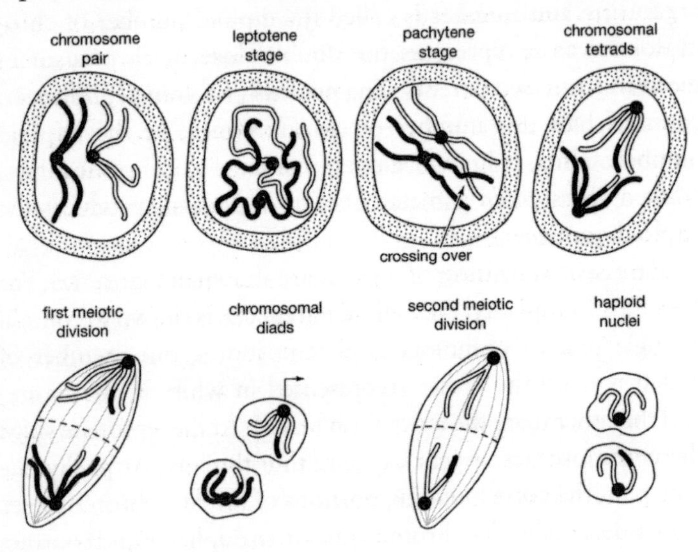

Figure 4.3 Meiosis.

Suppose that the dark chromosome carries the gene for albinism, and the light chromosome carries the gene for dark pigmentation. It is evident that the two gene alleles will undergo segregation at meiosis and that one-half of the gametes formed will contain the albino gene and the other half the pigmentation gene. Following the scheme in the diagram, random combination of the gametes with the albino gene and the pigmentation gene will give two kinds of homozygotes and one kind of heterozygote in a ratio of 1 : 1 : 2. Mendel's law of segregation is thus the outcome of chromosome behaviour at meiosis. The same is true of the second law, that of independent assortment.

Consider the inheritance of two pairs of genes, such as Mendel's factors for seed coloration and seed surface in peas;

these genes are located on different pairs of chromosomes. Since maternal and paternal members of different chromosome pairs are assorted independently, so are the genes they contain. This explains, in part, the genetic variety seen among the progeny of the same pair of parents. As stated above, humans have 46 chromosomes in the body cells and in the cells (oogonia and spermatogonia) from which the sex cells arise. At meiosis these 46 chromosomes form 23 pairs, one of the chromosomes of each pair being of maternal and the other of paternal origin. Independent assortment is, then, capable of producing 2^{23}, or 8,388,608, kinds of sex cells with different combinations of the grandmaternal and grandpaternal chromosomes. Since each parent has the potentiality of producing 2^{23} kinds of sex cells, the total number of possible combinations of the grandparental chromosomes is $2^{23} \times 2^{23} = 2^{46}$. The population of the world is now more than 6 billion, or approximately 2^{32}. It is therefore certain that only a tiny fraction of the potentially possible chromosome and gene combinations can ever be realized. Yet even 2^{46} is an underestimate of the variety potentially possible. The grandmaternal and grandpaternal members of the chromosome pairs are not indivisible units. Each chromosome carries many genes, and the chromosome pairs exchange segments at meiosis through the process of crossing over. This is evidence that the genes rather than the chromosomes are the units of Mendelian segregation.

Linkage of Traits

Simple linkage

As pointed out above, the random assortment of the maternal and paternal chromosomes at meiosis is the physical basis of the independent assortment of genes and of the traits they control. This is the basis of the second law of Mendel (see

Mendelian genetics, above). The number of the genes in a sex cell is, however, much greater than that of the chromosomes. When two or more genes are borne on the same chromosome, these genes may not be assorted independently; such genes are said to be linked. When a *Drosophila* fly homozygous for a normal grey body and long wings is crossed with one having a black body and vestigial wings, the F_1 consists of hybrid grey, long-winged flies. Grey body (*B*) is evidently dominant over black body (*b*), and long wing (*V*) is dominant over vestigial wing (*v*). Now consider a backcross of the heterozygous F_1 males to double-recessive black-vestigial females (*bbvv*). Independent assortment would be expected to give in the progeny of the backcross the following: 1 grey–long : 1 grey–vestigial : 1 black–long : 1 black–vestigial. In reality, only grey–long and black–vestigial flies are produced, in approximately equal numbers; the genes remain linked in the same combinations in which they were found in the parents. The backcross of the heterozygous F_1 females to double-recessive males gives a somewhat different result: 42 per cent each of grey–long and black–vestigial flies and about 8 per cent each of black–long and grey–vestigial classes. In sum, 84 per cent of the progeny have the parental combinations of traits, and 16 per cent have the traits recombined. The interpretation of these results given in 1911 by the American geneticist Thomas Hunt Morgan laid the foundation of the theory of linear arrangement of genes in the chromosomes.

Traits that exhibit linkage in experimental crosses (such as black body and vestigial wings) are determined by genes located in the same chromosome. As more and more genes became known in *Drosophila*, they fell neatly into four linkage groups corresponding to the four pairs of the chromosomes this species possesses. One linkage group consists of sex-linked genes, located in the X chromosome. Of the three remaining linkage groups, two have many more genes than the remaining one, which corresponds to the presence of two pairs of large chromosomes and one pair of tiny dot-like chromosomes. The numbers of linkage groups in other organisms are equal to or smaller than the numbers of the chromosomes in the sex cells – e.g. 10 linkage groups and 10 chromosomes in corn, 19 linkage groups and 20 chromosomes in the house mouse, and 23 linkage groups and 23 chromosomes in the human.

As described above, the linkage of the genes black and vestigial in *Drosophila* is complete in heterozygous males, while in the progeny of females there appear about 17 per cent of recombination classes. With very rare exceptions, the linkage of all genes belonging to the same linkage group is complete in *Drosophila* males, while in the females different pairs of genes exhibit all degrees of linkage from complete (no recombination) to 50 per cent (random assortment). Morgan's inference was that the degree of linkage depends on physical distance between the genes in the chromosome: the closer the genes, the tighter the linkage and vice versa. Furthermore, Morgan perceived that the chiasmata (crosses that occur in meiotic chromosomes) indicate the mechanism underlying the phenomena of linkage and crossing over. As shown schematically in Figure 4.3, the maternal and paternal chromosomes (represented as dark and light) cross over and exchange segments, so a chromosome emerging from the process of meiosis may consist of some maternal (grandmaternal) and some paternal (grandpaternal) sections. If the probability of crossing over taking place is uniform along the length of a chromosome (which was later shown to be not quite true), then genes close together will be recombined less frequently than those far apart.

This realization opened an opportunity to map the arrangement of the genes and the estimated distances between them in the chromosome by studying the frequencies of recombination of various traits in the progenies of hybrids. In other words, the linkage maps of the chromosomes are really summaries of many statistical observations on the outcomes of hybridization experiments. In principle at least, such maps could be prepared even if the chromosomes, not to speak of the chiasmata at meiosis, were unknown. But an interesting and relevant fact is that in *Drosophila* males the linkage of the genes in the same chromosome is complete, and observations under the microscope show that no chiasmata are formed in the chromosomes at meiosis. In most organisms, including humans, chiasmata are seen in the meiotic chromosomes in both sexes, and observations on hybrid progenies show that recombination of linked genes occurs also in both sexes.

Chromosome maps exist for the *Drosophila* fly, corn, the house mouse, the bread mould *Neurospora crassa*, and some bacteria and bacteriophages (viruses that infect bacteria). Until quite late in the 20th century, the mapping of human chromosomes presented a particularly difficult problem: experimental crosses could not be arranged in humans, and only a few linkages could be determined by analysis of unique family histories. However, the development of recombinant DNA technology provided new understanding of human genetic processes and new methods of research. Using the techniques of recombinant DNA technology, hundreds of genes have been mapped to the human chromosomes and many linkages established (see Chapter 9).

Sex linkage

The male of many animals has one chromosome pair, the sex chromosomes, consisting of unequal members called X and Y. At meiosis the X and Y chromosomes first pair then disjoin and pass to different cells. One-half of the gametes (spermatozoa) that are formed contain the X chromosome and the other half the Y. The female has two X chromosomes; all egg cells normally carry a single X. The eggs fertilized by X-bearing spermatozoa give females (XX), and those fertilized by Y-bearing spermatozoa give males (XY).

The genes located in the X chromosomes exhibit what is known as sex-linkage or criss-cross inheritance. This is because of a crucial difference between the paired sex chromosomes and the other pairs of chromosomes (called autosomes). The members of the autosome pairs are truly homologous; that is, each member of a pair contains a full complement of the same genes (albeit, perhaps, in different allelic forms). The sex chromosomes, on the other hand, do not constitute a homologous pair, as the X chromosome is much larger and carries far more genes than does the Y. Consequently, many recessive alleles carried on the X chromosome of a male will be expressed just as if they were dominant, for the Y chromosome carries no genes to counteract them.

The classic case of sex-linked inheritance, described by Morgan in 1910, is that of the white eyes in *Drosophila*. White-eyed females crossed to males with the normal red eye colour produce red-eyed daughters and white-eyed sons in the F_1 generation and equal numbers of white-eyed and red-eyed females and males in the F_2 generation. The cross of red-eyed females to white-eyed males gives a different result: both sexes are red-eyed in F_1 and the females in the F_2 generation are red-eyed, half the males are red-eyed, and the other half white-eyed. As interpreted by Morgan, the gene that determines the red or white eyes is borne on the X chromosome, and the allele for red eye is dominant over that for white eye. Since a male receives its single X chromosome from his mother, all sons of white-eyed females also have white eyes. A female inherits one X chromosome from her mother and the other X from her father. Red-eyed females may have genes for red eyes in both of their X chromosomes (homozygotes), or they may have one X with the gene for red and the other for white (heterozygotes). In the progeny of heterozygous females, one-half of the sons will receive the X chromosome with the gene for white and will have white eyes, and the other half will receive the X with the gene for red eyes. The daughters of the heterozygous females crossed with white-eyed males will have either two X chromosomes with the gene for white – and hence have white eyes – or one X with the gene for white and the other X with the gene for red and will be red-eyed heterozygotes.

In humans, red–green colour blindness and haemophilia are among many traits showing sex-linked inheritance and are consequently due to genes borne in the X chromosome.

In some animals – birds, butterflies and moths, some fish, and at least some amphibians and reptiles – the chromosomal mechanism of sex determination is a mirror image of that described above. The male has two X chromosomes and the female an X and Y chromosome. Here the spermatozoa all have an X chromosome; the eggs are of two kinds, some with X and others with Y chromosomes, usually in equal numbers. The sex of the offspring is then determined by the egg rather than by the spermatozoon. Sex-linked inheritance is altered correspondingly. A male homozygous for a sex-linked recessive trait crossed to a female with the dominant one gives, in the F_1 generation, daughters with the recessive trait and heterozygous sons with the corresponding dominant trait. The F_2 generation has recessive and dominant females and males in equal numbers. A male with a dominant trait crossed to a female with a recessive trait gives uniformly dominant F_1 and a segregation in a ratio of 2 dominant males : 1 dominant female : 1 recessive female.

Observations on pedigrees or experimental crosses show that certain traits exhibit sex-linked inheritance; the behaviour of the X chromosomes at meiosis is such that the genes they carry may be expected to exhibit sex-linkage. This evidence still failed to convince some sceptics that the genes for the sex-linked traits were in fact borne in certain chromosomes seen under the microscope. An experimental proof was furnished in 1916 by American geneticist Calvin Blackman Bridges. As stated above, white-eyed *Drosophila* females crossed to red-eyed males usually produce red-eyed female and white-eyed male progeny. Among thousands of such "regular" offspring, exceptional white-eyed females and red-eyed males are occasionally found. Bridges constructed the following working hypothesis. Suppose that, during meiosis in the female, gametogenesis occasionally goes wrong, and the two X chromosomes fail to disjoin. Exceptional eggs will then be produced, some carrying two X chromosomes and some carrying none. An egg with two X chromosomes coming from a white-eyed female fertilized by a spermatozoon with a Y chromosome will give an exceptional white-eyed female. An egg with no X chromosome fertilized by a spermatozoon with an X chromosome derived from a red-eyed father will yield an exceptional red-eyed male. This hypothesis can be rigorously tested. The exceptional white-eyed females should have not only the two X chromosomes but also a Y chromosome, which normal females do not have. The exceptional males should, on the other hand, lack a Y chromosome, which normal males do have. Both predictions were verified by examination under a microscope of the chromosomes of exceptional females and males. The hypothesis also predicts that exceptional eggs with two X chromosomes fertilized by X-bearing spermatozoa must give individuals with three X chromosomes; such individuals were later identified by Bridges as poorly viable "superfemales". Exceptional eggs with no Xs, fertilized by Y-bearing spermatozoa, will give zygotes without X chromosomes; such zygotes die in early stages of development.

Chromosomal Aberrations

The chromosome set of a species remains relatively stable over long periods of time. However, within populations there can be found abnormalities involving the structure or number of chromosomes. These alterations arise spontaneously from errors in the normal processes of the cell. Their consequences are usually deleterious, giving rise to individuals who are unhealthy or sterile, though in rare cases alterations provide new adaptive opportunities that allow evolutionary change to occur. In fact, the discovery of visible chromosomal differences between species has given rise to the belief that radical re-structuring of chromosome architecture has been an important force in evolution.

Changes in chromosome structure

Two important principles dictate the properties of a large proportion of structural chromosomal changes. The first principle is that any deviation from the normal ratio of genetic material in the genome results in genetic imbalance and abnormal function. In the normal nuclei of both diploid and haploid cells, the ratio of the individual chromosomes to one another is 1 : 1. Any deviation from this ratio by addition or subtraction of either whole chromosomes or parts of chromosomes results in genomic imbalance. The second principle is that homologous chromosomes go to great lengths to pair at meiosis. The tightly paired homologous regions are joined by a ladder-like longitudinal structure called the syn-aptonemal complex. Homologous regions seem to be able to find each other and form a synaptonemal complex whether or not they are part of normal chromosomes. Therefore, when structural changes occur, not only are the resulting pairing formations highly characteristic of that type of structural change but they also dictate the packaging of normal and abnormal chromosomes into the gametes and subsequently into the progeny.

Deletions

The simplest, but perhaps most damaging, structural change is a deletion – the complete loss of a part of one chromosome. In a haploid cell this is lethal, because part of the essential genome is lost. However, even in diploid cells deletions are generally lethal or have other serious consequences. In a diploid a heterozygous deletion results in a cell that has one normal chromosome set and another set that contains a truncated chromosome. Such cells show genomic imbalance, which increases in severity with the size of the deletion. Another potential source of damage is that any recessive, deleterious, or lethal alleles that are in the normal counterpart of the deleted region will be expressed in the phenotype. In humans, *cri-du-chat* (French for "cry of the cat") syndrome is caused by a heterozygous deletion at the tip of the short arm of chromosome 5. Infants are born with this condition as the result of a deletion arising in parental germinal tissues or even in sex cells. The manifestations of this deletion, in addition to the "cat cry" that gives the syndrome its name, include severe intellectual disability and an abnormally small head.

Duplications

A heterozygous duplication (an extra copy of some chromosome region) also results in a genomic imbalance with deleterious consequences. Small duplications within a gene can arise spontaneously. Larger duplications can be caused by crossovers following asymmetrical chromosome pairing or by meiotic irregularities resulting from other types of altered chromosome structures. If a duplication becomes homozygous, it can provide the organism with an opportunity to acquire new genetic functions through mutations within the duplicate copy.

Inversions

An inversion occurs when a chromosome breaks in two places and the region between the break rotates 180° before rejoining with the two end fragments. If the inverted segment contains the centromere (i.e. the point where the two chromatids are joined), the inversion is said to be pericentric; if not, it is called paracentric. Inversions do not result in a gain or loss of genetic material, and they have deleterious effects only if one of the chromosomal breaks occurs within an essential gene or if the function of a gene is altered by its relocation to a new chromosomal neighbourhood (called the position effect). However, individuals who are heterozygous for inversions produce aberrant meiotic products along with normal products. The only way uninverted and inverted segments can pair is by forming an inversion loop. If no crossovers occur in the loop, half of the gametes will be normal and the other half will contain an inverted chromosome. If a crossover does occur within the loop of a paracentric inversion, a chromosome bridge and an acentric chromosome (i.e. a chromosome without a centromere) will be formed, and this will give rise to abnormal meiotic products carrying deletions, which are not viable. In a pericentric inversion, a crossover within the loop does not result in a bridge or an acentric chromosome, but non-viable products are produced carrying a duplication and a deletion.

Translocations

If a chromosome break occurs in each of two non-homologous chromosomes and the two breaks rejoin in a new arrangement, the new segment is called a translocation. A cell bearing a heterozygous translocation has a full set of genes and will be viable unless one of the breaks causes damage within a gene or if there is a position effect on gene function. However, once again the pairing properties of the chromosomes at meiosis result in aberrant meiotic products. Specifically, half of the products are deleted for one of the chromosome regions that changed positions and half of the products are duplicated for the other. These duplications and deletions usually result in inviability, so translocation heterozygotes are generally semi-sterile ("half-sterile").

Changes in chromosome number

Two types of changes in chromosome numbers can be distinguished: a change in the number of whole chromosome sets (polyploidy) and a change in chromosomes within a set (aneuploidy).

Polyploids

An individual with additional chromosome sets is called a polyploid. Individuals with three sets of chromosomes (triploids, $3n$) or four sets of chromosomes (tetraploids, $4n$) are polyploid derivatives of the basic diploid ($2n$) constitution. Polyploids with odd numbers of sets (e.g. triploids) are sterile, because homologous chromosomes pair only two by two, and the extra chromosome moves randomly to a cell pole, resulting in highly unbalanced, non-functional meiotic products. It is for this reason that triploid watermelons are seedless. However, polyploids with even numbers of chromosome sets can be fertile if orderly, two-by-two chromosome pairing occurs.

Though two organisms from closely related species frequently hybridize, the chromosomes of the fusing partners are different enough that the two sets do not pair at meiosis, resulting in sterile offspring. However, if by chance the number of chromosome sets in the hybrid accidentally duplicates, a pairing partner for each chromosome will be produced, and the hybrid will be fertile. These chromosomally doubled hybrids are called allotetraploids. Bread wheat, which is hexaploid ($6n$), as a result of several natural spontaneous hybridizations, is an example of an allotetraploid. Some polyploid plants are able to produce seeds through an asexual type of reproduction called apomixis; in such cases, all progeny are identical to the parent. Polyploidy does arise spontaneously in humans, but all polyploids either abort in utero or die shortly after birth. (See also Chapter 5, Polyploidy, where the importance of polyploids to evolution is discussed in more detail.)

Aneuploids

Some cells have an abnormal number of chromosomes that is not a whole multiple of the haploid number. This condition is called aneuploidy. Most aneuploids arise by non-disjunction, a failure of homologous chromosomes to separate at meiosis. When a gamete of this type is fertilized by a normal gamete, the zygotes formed will have an unequal distribution of chromosomes. Such genomic imbalance results in severe abnormalities or death. Only aneuploids involving small chromosomes tend to survive and even then only with an aberrant phenotype.

The most common form of aneuploidy in humans results in Down syndrome, a suite of specific disorders in individuals possessing an extra chromosome 21 (trisomy 21). The symptoms of Down syndrome include intellectual disability, severe disorders of internal organs such as the heart and kidneys, upslanted eyes, an enlarged tongue, and abnormal dermal ridge patterns on the fingers, palms, and soles.

Other forms of aneuploidy in humans result from abnormal numbers of sex chromosomes. Turner syndrome is a condition in which females have only one X chromosome. Symptoms may include short stature, webbed neck, kidney or heart malformations, underdeveloped sex characteristics, or sterility. Klinefelter syndrome is a condition in which males have one extra female sex chromosome, resulting in an XXY pattern. (Other, less frequent, chromosomal patterns include XXXY, XXXXY, XXYY, and XXXYY.) Symptoms of Klinefelter syndrome may include sterility, a tall physique, lack of secondary sex characteristics, breast development, and learning disabilities.

Human Chromosomes

Fertilization, Sex Determination, and Differentiation

A human individual arises through the union of two cells, an egg from the mother and a sperm from the father. Human egg cells are barely visible to the naked eye. They are shed, usually one at a time, from the ovary into the oviducts (fallopian tubes), through which they pass into the uterus. Fertilization, the penetrance of an egg by a sperm, occurs in the oviducts. This is the main event of sexual reproduction and determines the genetic constitution of the new individual.

Human sex determination is a genetic process that depends basically on the presence of the Y chromosome in the fertilized egg. This chromosome stimulates a change in the undifferentiated gonad into that of the male (a testicle). The gonadal action of the Y chromosome is mediated by a gene located near the centromere; this gene codes for the production of a cell surface molecule called the H-Y antigen. Further development of the anatomic structures, both internal and external, that are associated with maleness is controlled by hormones produced by the testicle. The sex of an individual can be thought of in three different contexts: chromosomal sex, gonadal sex, and anatomic sex. Discrepancies among these, especially the latter two, result in the development of individuals with ambiguous sex, often called hermaphrodites. (This is unrelated to homosexuality, the cause of which is uncertain.)

It is of interest that in the absence of a male gonad (testicle) the internal and external sexual anatomy is always female, even in the absence of a female ovary. A female without ovaries will, of course, be infertile and will not experience any of the female developmental changes normally associated with puberty. Such females often have Turner syndrome.

If X-containing and Y-containing sperm are produced in equal numbers, then according to simple chance one would expect the sex ratio at conception (fertilization) to be half boys and half girls, or 1 : 1. Direct observation of sex ratios among newly fertilized human eggs is not yet feasible, and sex-ratio data are usually collected at the time of birth. In almost all human populations of newborns there is a slight excess of males; about 106 boys are born for each 100 girls. Throughout life, however, there is a slightly greater mortality of males; this slowly alters the sex ratio until, beyond the age of about 50 years, there is an excess of females. Studies indicate that male embryos suffer a relatively greater degree of prenatal mortality, so that the sex ratio at conception might be expected to favour males even more than the 106 : 100 ratio observed at birth would suggest. Firm explanations for the apparent excess of male conceptions have not been established; it is possible that Y-containing sperm survive better within the female reproductive tract, or that they may be a little more successful in reaching the egg in order to fertilize it. In any case, the sex differences are small, the statistical expectation for a boy (or girl) at any single birth still being close to 0.5 (one out of two).

During gestation – the period of nine months between fertilization and the birth of the infant – a remarkable series of developmental changes occur. Through the process of mitosis, the total number of cells changes from 1 (the fertilized egg) to about 2×10^{11}. In addition, these cells differentiate into hundreds of different types with specific functions (liver cells, nerve cells, muscle cells, etc.). A multitude of regulatory processes, both genetically and environmentally controlled, accomplish this differentiation. Elucidation of the exquisite timing of these processes remains one of the great challenges of human biology.

Microevolution and Human Evolution

At the centre of the theory of evolution as proposed by Charles Darwin and Alfred Russel Wallace were the concepts of variation and natural selection. Hereditary variants were thought to arise naturally in populations, and then these were selected either for or against by the contemporary environmental conditions. In this way, subsequent generations became either enriched or impoverished for specific variant types. Over the long term, the accumulation of such changes in populations could lead to the formation of new species and higher taxonomic categories. However, although hereditary change was basic to the theory, in the 19th-century world of Darwin and Wallace, the fundamental unit of heredity – the gene – was unknown. The birth and proliferation of the science of genetics in the 20th century after the discovery of Mendel's laws made it possible to consider the process of evolution by natural selection in terms of known genetic processes.

Microevolution

Even within the relatively short period of time since Darwin, it has been possible to document small-scale evolution, or microevolution. Allelic variation has been found to be common in nature. It is detected as polymorphism, the presence of two or more distinct hereditary forms associated with a gene. Polymorphism can be morphological, such as blue and brown forms of a species of marine mussel, or molecular, detectable only at the DNA or protein level. Although much of this polymorphism is not understood, there are enough examples of selection of polymorphic forms to indicate that it is potentially adaptive. Selection has been observed favouring melanic (dark) forms of peppered moths in industrial areas and favouring resistance to toxic agents such as the insecticide DDT, the rat poison warfarin, and the virus that causes the disease myxomatosis in rabbits.

More complex genetic changes have been documented, leading to special locally adapted "ecotypes". Anoles (a type of lizard) on certain Caribbean islands show convincing examples of adaptations to specific habitats, such as tree trunks, tree branches, or grass. Introductions of lizards to uncolonized islands result in demonstrable microevolutionary adaptations to the various vacant niches. In the Galapagos, studies over several decades have documented adaptive changes in the beaks of finches (see Figure 5.3). In some studies, documented changes have led to incipient new species. An example is the apple maggot, the larva of a fly in North America that has evolved from a similar fly living on hawthorns – all in the period since the introduction of apples by European settlers.

The formation of new species was a key component of Darwin's original theory. Now it appears that the accumulation of enough small-scale genetic changes can lead to the inability to mate with members of an ancestral population. Such reproductive isolation is the key step in species formation.

It is reasonable to assume that the continuation of microevolutionary genetic changes over very long periods of time can give rise to new major taxonomic groups, the process of macroevolution. There are few data that bear directly on the processes of macroevolution, but gene analysis does provide a way for charting macroevolutionary relationships indirectly, as shown below.

DNA phylogeny

The ability to isolate and sequence specific genes and genomes has been of great significance in deducing trees of evolutionary relatedness. An important discovery that enables this sort of analysis is the considerable evolutionary conservation between organisms at the genetic level. This means that different organisms have a large proportion of their genes in common, particularly those that code for proteins at the central core of the chemical machinery of the cell. For example, most organisms have a gene coding for the energy-producing protein cytochrome c, and furthermore, this gene has a very similar nucleotide sequence in all organisms (that is, the sequence is conserved). However, the sequences of cytochrome c in different organisms do show differences, and the key to phylogeny is that the differences are proportionately fewer between organisms that are closely related. The interpretation of this observation is that organisms that share a common ancestor also share common DNA sequences derived from that ancestor. When one ancestral species splits into two, differences accumulate as a result of mutations, a process called divergence. The greater the amount of divergence, the longer must have been the time since the split occurred. To carry out this sort of analysis, the DNA sequence data are fed into a computer. The computer positions similar species together on short adjacent branches showing a relatively recent split and dissimilar species on long branches from an ancient split. In this way a molecular phylogenetic tree of any number of organisms can be drawn (Figure 4.4).

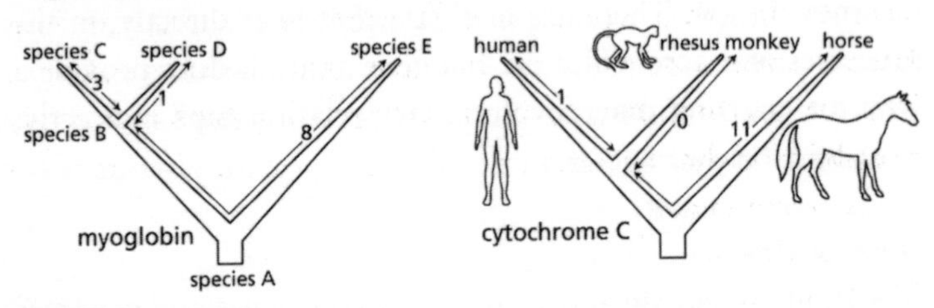

Figure 4.4 DNA phylogeny. (Left) Amount of change in the evolutionary history of three hypothetical living species (C, D, and E), inferred by comparing amino acid differences in their myoglobin molecules. All three species have the same earlier ancestor (A). (Right) Phylogeny of the human, the rhesus monkey, and the horse, based on amino-acid substitutions in the evolution of cytochrome c in the lineages of the three species.

DNA difference in some cases can be correlated with absolute dates of divergence as deduced from the fossil record. Then it is possible to calculate divergence as a rate. It has been found that divergence is relatively constant in rate, giving rise to the idea that there is a type of "molecular clock" ticking in the course of evolution. Some ticks of this clock (in the form of mutations) are significant in terms of adaptive changes to the gene, but many are undoubtedly neutral, with no significant effect on fitness.

One of the interesting discoveries to emerge from molecular phylogeny is that gene duplication has been common during evolution. If an extra copy of a gene can be made, initially by some cellular accident, then the "spare" copy is free to mutate and evolve into a separate function.

Molecular phylogeny of some genes has also pointed to unexpected cases of, say, a plant gene nested within a tree of animal genes of that type or a bacterial gene nested within a plant phylogenetic tree. The explanation for such anomalies is that there has been horizontal transmission from one group to another. In other words, on rare occasions a gene can hop laterally from one species to another. Although the mechanisms for horizontal transmission are presently not known, one possibility is that bacteria or viruses act as natural vectors for transferring genes.

Synteny

Genomic sequencing and mapping have enabled comparison of the general structures of genomes of many different species. The general finding is that organisms of relatively recent divergence show similar blocks of genes in the same relative positions in the genome. This situation is called synteny, translated roughly as possessing common chromosome sequences. For example, many of the genes of humans are syntenic with those of other mammals – not only apes but also cows, mice, and so on. Study of synteny can show how the genome is "cut and pasted" in the course of evolution.

Human Evolution

Many of the techniques of evolutionary genetics can be applied to the evolution of humans. Charles Darwin created a great controversy in Victorian Britain by suggesting in his book *The Descent of Man* that humans and apes share a common ancestor. Darwin's assertion was based on the many shared anatomical features of apes and humans. DNA analysis has supported this hypothesis. At the DNA sequence level, the genomes of humans and chimpanzees are 99 per cent identical.

Furthermore, when phylogenetic trees are constructed using individual genes, humans and apes cluster together in short terminal branches of the trees, suggesting very recent divergence. Synteny too is impressive, with relatively minor chromosomal rearrangements.

Fossils have been found of various extinct forms considered to be intermediates between apes and humans. Notable is the African genus *Australopithecus*, generally believed to be one of the earliest hominins and an intermediate on the path of human evolution. The first toolmaker was *Homo habilis*, followed by *Homo erectus*, and finally *Homo sapiens* (modern humans). *H. habilis* fossils have been found only in Africa, whereas fossils of *H. erectus* and *H. sapiens* are found throughout the Old World. Phylogenetic trees based on DNA sequencing of people thoughout the world have shown that Africans represent the root of the trees. This is interpreted as evidence that *H. sapiens* evolved in Africa, spread throughout the globe, and outcompeted *H. erectus* wherever the two cohabited.

Variations of DNA, either unique alleles of individual genes or larger-sized blocks of variable structure, have been used as markers to trace human migrations across the globe. Hence, it has been possible to trace the movement of *H. sapiens* out of Africa and into Europe and Asia and, more recently, to the American continents. Also, genetic markers are useful in plotting human migrations that occurred in historical time. For example, the invasion of Europe by various Asian conquerors can be followed using blood-type alleles.

As humans colonized and settled permanently in various parts of the world, they differentiated themselves into distinct groups called races. Undoubtedly, many of the features that distinguish races, such as skin colour or body shape, were adaptive in the local settings, although such adaptiveness is difficult to demonstrate. Nevertheless, genomic analysis has revealed that the concept of race has little meaning at the genetic level. The differences between races are superficial, based on the alleles of a relatively small number of genes that affect external features. Furthermore, while races differ in allele frequencies, these same alleles are found in most races. In other words, at the genetic level there are no significant discontinuities between races. It is paradoxical that race, which has been so important to people throughout the course of human history, is trivial at the genetic level – an important insight to emerge from genetic analysis.

5

EVIDENCE FOR EVOLUTION

Introduction

Darwin and other 19th-century biologists found compelling evidence for biological evolution in the comparative study of living organisms, in their geographic distribution, and in the fossil remains of extinct organisms. Since Darwin's time, the evidence from these sources has become considerably stronger and more comprehensive, while biological disciplines that have emerged more recently – genetics, biochemistry, physiology, ecology, animal behaviour (ethology), and especially molecular biology – have supplied powerful additional evidence and detailed confirmation. The amount of information about evolutionary history stored in the DNA and proteins of living things is virtually unlimited; scientists can reconstruct any detail of the evolutionary history of life by investing sufficient time and laboratory resources.

Evolutionists are no longer concerned with obtaining evidence to support the fact of evolution but rather are concerned with what sorts of knowledge can be obtained from different sources of evidence. The most productive of these sources are the fossil record, anatomical similarities between organisms, embryonic development and vestiges, biogeography, and molecular biology. Only the last of these is discussed in this section, as the others are beyond the scope of this book.

Molecular Biology

The field of molecular biology provides the most detailed and convincing evidence available for biological evolution. In its unveiling of the nature of DNA and the workings of organisms at the level of enzymes and other protein molecules, it has shown that these molecules hold information about an organism's ancestry. This has made it possible to reconstruct evolutionary events that were previously unknown and to confirm and adjust the view of events already known. The precision with which these events can be reconstructed is one reason the evidence from molecular biology is so compelling. Another reason is that molecular evolution has shown all living organisms, from bacteria to humans, to be related by descent from common ancestors.

A remarkable uniformity exists in the molecular components of organisms – in the nature of the components as well as in the ways in which they are assembled and used. In all bacteria, plants, animals, and humans, the DNA comprises a different sequence of the same 4 component nucleotides, and all the various proteins are synthesized from different combinations and sequences of the same 20 amino acids, although several hundred other amino acids exist. The genetic code by which the information contained in the DNA of the cell nucleus is passed on to proteins is virtually everywhere the same. Similar metabolic pathways – sequences of biochemical reactions – are used by the most diverse organisms to produce energy and to make up the cell components.

This unity reveals the genetic continuity and common ancestry of all organisms. There is no other rational way to account for their molecular uniformity when numerous alternative structures are equally likely. The genetic code serves as an example. Each particular sequence of three nucleotides in the nuclear DNA acts as a pattern for the production of exactly the same amino acid in all organisms. This is no more necessary than it is for a language to use a particular combination of letters to represent a particular object. If it is found that certain sequences of letters – *planet*, *tree*, *woman* – are used with identical meanings in a number of different books, one can be sure that the languages used in those books are of common origin.

Genes and proteins are long molecules that contain information in the sequence of their components in much the same way as sentences of the English language contain information in the sequence of their letters and words. The sequences that make up the genes are passed on from parents to offspring and are identical except for occasional changes introduced by mutations. As an illustration, assume that two books are being compared. Both books are 200 pages long and contain the same number of chapters. Closer examination reveals that the two books are identical page for page and word for word, except that an occasional word – say, 1 in 100 – is different. The two books cannot have been written independently; either one has been copied from the other, or both have been copied, directly or indirectly, from the same original book. Similarly, if each component nucleotide of DNA is represented by one letter, the complete sequence of nucleotides in the DNA of a higher organism would require several hundred books of hundreds of pages, with several thousand letters on each page.

When the "pages" (or sequences of nucleotides) in these "books" (organisms) are examined one by one, the correspondence in the "letters" (nucleotides) gives unmistakable evidence of common origin.

The two arguments presented above are based on different grounds, although both attest to evolution. Using the alphabet analogy, the first argument says that languages that use the same dictionary – the same genetic code and the same 20 amino acids – cannot be of independent origin. The second argument, concerning similarity in the sequence of nucleotides in the DNA (and thus the sequence of amino acids in the proteins), says that books with very similar texts cannot be of independent origin.

The evidence of evolution revealed by molecular biology goes even farther. The degree of similarity in the sequence of nucleotides or of amino acids can be precisely quantified. For example, in humans and chimpanzees, the protein molecule called cytochrome c, which serves a vital function in respiration within cells, consists of the same 104 amino acids in exactly the same order. It differs, however, from the cytochrome c of rhesus monkeys by 1 amino acid, from that of horses by 11 additional amino acids, and from that of tuna by 21 additional amino acids. The degree of similarity reflects the recency of common ancestry. Thus, inferences from comparative anatomy and other disciplines concerning evolutionary history can be tested in molecular studies of DNA and proteins by examining their sequences of nucleotides and amino acids. (See below, DNA and Protein as Informational Macromolecules.)

The authority of this kind of test is overwhelming; each of the thousands of genes and thousands of proteins contained in an organism provides an independent test of that organism's evolutionary history. Not all possible tests have been performed, but many hundreds have been done, and not one has given evidence contrary to evolution. There is probably no other notion in any field of science that has been as extensively tested and as thoroughly corroborated as the evolutionary origin of living organisms.

Process of Evolution

Evolution as a Genetic Function

The concept of natural selection

The central argument of Darwin's theory of evolution starts with the existence of hereditary variation. Experience with animal and plant breeding had demonstrated to Darwin that variations can be developed that are "useful to man". So, he reasoned, variations must occur in nature that are favourable or useful in some way to the organism itself in the struggle for existence. Favourable variations are ones that increase chances for survival and procreation. Those advantageous variations are preserved and multiplied from generation to generation at the expense of less-advantageous ones. This is the process known as natural selection. The outcome of the process is an organism that is well adapted to its environment, and evolution often occurs as a consequence.

Natural selection, then, can be defined as the differential reproduction of alternative hereditary variants, determined by the fact that some variants increase the likelihood that the organisms having them will survive and reproduce more successfully than will organisms carrying alternative variants. Selection may occur as a result of differences in survival, in fertility, in rate of development, in mating success, or in any other aspect of the life cycle. All of these differences can be incorporated under the term "differential reproduction" because all result in natural selection to the extent that they affect the number of progeny an organism leaves.

Darwin maintained that competition for limited resources results in the survival of the most effective competitors. Nevertheless, natural selection may occur not only as a result of competition but also as a result of some aspect of the physical environment, such as inclement weather. Moreover, natural selection would occur even if all the members of a population died at the same age, simply because some of them would have produced more offspring than others. Natural selection is quantified by a measure called Darwinian fitness or relative fitness. Fitness in this sense is the relative probability that a hereditary characteristic will be reproduced; that is, the degree of fitness is a measure of the reproductive efficiency of the characteristic.

Biological evolution is the process of change and diversification of living things over time, and it affects all aspects of their lives – morphology (form and structure), physiology, behaviour, and ecology. Underlying these changes are changes in the hereditary materials. Hence, in genetic terms evolution consists of changes in the organism's hereditary make-up.

Evolution can be seen as a two-step process. First, hereditary variation takes place; second, selection is made of those genetic variants that will be passed on most effectively to the following generations. Hereditary variation also entails two mechanisms – the spontaneous mutation of one variant into another and the sexual process that recombines those variants (recombination) to form a multitude of variations. The variants that arise by mutation or recombination are not transmitted equally from one generation to another. Some may appear more frequently because they are favourable to the organism; the frequency of others may be determined by accidents of chance, called genetic drift.

Genetic variation in populations

The gene pool

The gene pool is the sum total of all the genes and combinations of genes that occur in a population of organisms of the same species. It can be described by citing the frequencies of the alternative genetic constitutions. Consider, for example, a particular gene (which geneticists call a locus), such as the one determining the MN blood groups in humans. One form of the gene codes for the M blood group, while the other form codes for the N blood group; different forms of the same gene are called alleles. The MN gene pool of a particular population is specified by giving the frequencies of the alleles *M* and *N*. Thus, in the United States the *M* allele occurs in people of European descent with a frequency of 0.539 and the *N* allele with a frequency of 0.461 – that is, 53.9 per cent of the alleles in the population are *M* and 46.1 per cent are *N*. In other populations these frequencies are different; for instance, the frequency of the *M* allele is 0.917 in Navajo Indians and 0.178 in Australian Aboriginals.

The necessity of hereditary variation for evolutionary change to occur can be understood in terms of the gene pool. Assume, for instance, a population in which there is no variation at the gene locus that codes for the MN blood groups; only the *M* allele exists in all individuals. Evolution of the MN blood groups cannot take place in such a population, since the allelic frequencies have no opportunity to change from generation to generation. On the other hand, in populations in which both alleles *M* and *N* are present, evolutionary change is possible.

Genetic variation and rate of evolution

The more genetic variation that exists in a population, the greater the opportunity for evolution to occur. As the number of gene loci that are variable increases and as the number of alleles at each locus becomes greater, the likelihood grows that some alleles will change in frequency at the expense of their alternates. The British geneticist R.A. Fisher mathematically demonstrated a direct correlation between the amount of genetic variation in a population and the rate of evolutionary change by natural selection. This demonstration is embodied in his fundamental theorem of natural selection (1930): "The rate of increase in fitness of any organism at any time is equal to its genetic variance in fitness at that time."

This theorem has been confirmed experimentally. One study employed different strains of *Drosophila serrata*, a species of vinegar fly from eastern Australia and New Guinea. Evolution in these flies can be investigated by breeding them in separate "population cages" and finding out how populations change over many generations. Experimental populations were set up, with the flies living and reproducing in their isolated microcosms. Single-strain populations were established from flies collected either in New Guinea or in Australia; in addition, a mixed population was constituted by crossing these two strains of flies. The mixed population had the greater initial genetic variation, since it began with two different single-strain populations. To encourage rapid evolutionary change, the populations were manipulated such that the flies experienced intense competition for food and space. Adaptation to the experimental environment was measured by periodically counting the number of individuals in the populations.

Two results deserve notice. First, the mixed population had, at the end of the experiment, more flies than the single-strain populations. Second, and more relevant, the number of flies increased at a faster rate in the mixed population than in the single-strain populations. Evolutionary adaptation to the environment occurred in both types of population; both were able to maintain higher numbers as the generations progressed. But the rate of evolution was more rapid in the mixed group than in the single-strain groups. The greater initial amount of genetic variation made possible a faster rate of evolution.

Measuring gene variability

Because a population's potential for evolving is determined by its genetic variation, evolutionists are interested in discovering the extent of such variation in natural populations. It is readily apparent that plant and animal species are heterogeneous in all sorts of ways – in the flower colours and growth habits of plants, for instance, or the shell shapes and banding patterns of snails. Differences are more readily noticed among humans – in facial features, hair and skin colour, height, and weight – but such morphological differences are present in all groups of organisms. One problem with morphological variation is that it is not known how much is due to genetic factors and how much may result from environmental influences.

Animal and plant breeders select for their experiments individuals or seeds that excel in desired attributes – the protein content of corn (maize), for example, or the milk yield of cows. The selection is repeated generation after generation. If the population changes in the direction favoured by the breeder, it becomes clear that the original stock possessed genetic variation with respect to the selected trait.

The results of artificial selection are impressive. Selection for high oil content in corn increased the oil content from less than 5 per cent to more than 19 per cent in 76 generations, while selection for low oil content reduced it to below 1 per cent. Thirty years of selection for increased egg production in a flock of White Leghorn chickens increased the average yearly output of a hen from 125.6 to 249.6 eggs. Artificial selection has produced endless varieties of dog, cat, and horse breeds. The plants grown for food and fibre and the animals bred for food and transportation are all products of age-old or modern-day artificial selection.

Since the late 20th century, scientists have used the techniques of molecular biology to modify or introduce genes for desired traits in a variety of organisms, including domestic plants and animals; this field has become known as genetic engineering or recombinant DNA technology. Improvements that in the past were achieved after tens of generations by artificial selection can now be accomplished much more effectively and rapidly (within a single generation) by molecular genetic technology.

The success of artificial selection for virtually every trait and every organism in which it has been tried suggests that genetic variation is pervasive throughout natural populations. But evolutionists like to go one step further and obtain quantitative estimates. Only since the 1960s, with the advances of molecular biology, have geneticists developed methods for measuring the extent of genetic variation in populations or among species of organisms. These methods consist essentially of taking a sample of genes and finding out how many are variable and how variable each one is. One simple way of measuring the variability of a gene locus is to ascertain what proportion of the individuals in a population are heterozygotes at that locus. In a heterozygous individual the two genes for a trait, one received from the mother and the other from the father, are different. The proportion of heterozygotes in the population is, therefore, the same as the probability that two genes taken at random from the gene pool are different.

Techniques for determining heterozygosity have been used to investigate numerous species of plants and animals. Typically, insects and other invertebrates are more varied genetically than mammals and other vertebrates, and plants bred by outcrossing (crossing with relatively unrelated strains) exhibit more variation than those bred by self-pollination. But the amount of genetic variation is in any case astounding. Consider as an example humans, whose level of variation is about the same as that of other mammals. The human heterozygosity value at the level of proteins is stated as $H = 0.067$, which means that an individual is heterozygous at 6.7 per cent of his or her genes, because the two genes at each locus encode slightly different proteins. The Human Genome Project demonstrated that there are approximately 25,000 genes in humans. This means that a person is heterozygous at no fewer than $25{,}000 \times 0.067 = 1{,}675$ gene loci. An individual heterozygous at one locus (*Aa*) can produce two different kinds of sex cells, or gametes, one with each allele (*A* and *a*); an individual heterozygous at two loci (*AaBb*) can produce four kinds of gametes (*AB*, *Ab*, *aB*, and *ab*); an individual heterozygous at *n* loci can potentially produce 2^n different gametes. Therefore, a typical human individual has the potential to produce 2^{1675}, or approximately 10^{502} (1 with 502 zeros following), different kinds of gametes. That number is much larger than the estimated number of atoms in the universe, about 10^{80}.

It is clear, then, that every sex cell produced by a human being is genetically different from every other sex cell and, therefore, that no two persons who ever existed or will ever exist are likely to be genetically identical – with the exception of identical twins, which develop from a single fertilized ovum. The same conclusion applies to all organisms that reproduce sexually; every individual represents a unique genetic configuration that will likely never be repeated again. This enormous reservoir of genetic variation in natural populations provides virtually unlimited opportunities for evolutionary change in response to the environmental constraints and the needs of the organisms.

Origin of Genetic Variation: Mutations

Life originated about 3.5 billion years ago in the form of primordial organisms that were relatively simple and very small. All living things have evolved from these lowly beginnings. At present there are more than 2 million known species, which are widely diverse in size, shape, and way of life, as well as in the DNA sequences that contain their genetic information. What has produced the pervasive genetic variation within natural populations and the genetic differences among species? There must be some evolutionary means by which existing DNA sequences are changed and new sequences are incorporated into the gene pools of species.

The information encoded in the nucleotide sequence of DNA is, as a rule, faithfully reproduced during replication, so that each replication results in two DNA molecules that are identical to each other and to the parent molecule. But heredity is not a perfectly conservative process; otherwise, evolution could not have taken place. Occasionally "mistakes", or mutations, occur in the DNA molecule during replication, so that daughter cells differ from the parent cells in the sequence or in the amount of DNA. A mutation first appears in a single cell of an organism, but it is passed on to all cells descended from the first. Mutations can be classified into two categories – gene, or point, mutations, which affect only a few nucleotides within a gene, and chromosomal mutations, which either change the number of chromosomes or change the number or arrangement of genes on a chromosome.

Gene mutations

A gene mutation occurs when the nucleotide sequence of the DNA is altered and a new sequence is passed on to the offspring. The change may be either a substitution of one or a few nucleotides for others or an insertion or deletion of one or a few pairs of nucleotides.

The four nucleotide bases of DNA are represented by the letters A, C, G, and T (see Chapter 1, Structure and Composition of DNA). A gene that bears the code for constructing a protein molecule consists of a sequence of several thousand nucleotides, so that each segment of three nucleotides – called a triplet or codon – codes for one particular amino acid in the protein. The nucleotide sequence in the DNA is first transcribed into a molecule of messenger RNA (mRNA). The mRNA, using a slightly different code (represented by the letters A, C, G, and U), bears the message that determines which amino acid will be inserted into the protein's chain in the process of translation. Substitutions in the nucleotide sequence of a structural gene may result in changes in the amino acid sequence of the protein, although this is not always the case. The genetic code is redundant in that different triplets may hold the code for the same amino acid. Consider the triplet AUA in messenger RNA, which codes for the amino acid isoleucine. If the last A is replaced by C, the triplet still codes for isoleucine, but if it is replaced by G, it codes for methionine instead (Figure 5.1).

A nucleotide substitution in the DNA that results in an amino acid substitution in the corresponding protein may or may not severely affect the biological function of the protein. Some nucleotide substitutions change a codon for an amino acid into a signal to terminate translation, and those mutations are likely to have harmful effects. If, for instance, the second U in the triplet UUA, which codes for leucine, is replaced by A, the triplet becomes UAA, a "terminator" codon; the result is that the triplets following this codon in the DNA sequence are not translated into amino acids.

Additions or deletions of nucleotides within the DNA sequence of a structural gene often result in a greatly altered

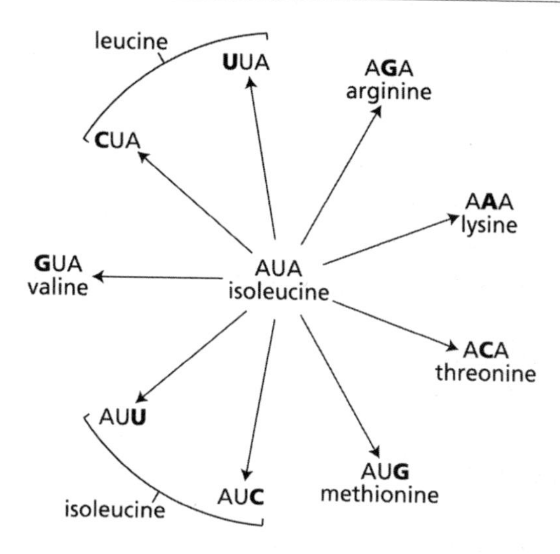

Figure 5.1 Effect of nucleotide substitutions on codons for amino acids. The effect of base substitutions, or point mutations, on the messenger-RNA codon AUA, which codes for the amino acid isoleucine. Substitutions (**bold** letters) at the first, second, or third position in the codon can result in nine new codons corresponding to six different amino acids in addition to isoleucine itself. The chemical properties of some of these amino acids are quite different from those of isoleucine. Replacement of one amino acid in a protein by another can seriously affect the protein's biological function.

sequence of amino acids in the coded protein. The addition or deletion of one or two nucleotides shifts the "reading frame" of the nucleotide sequence all the way from the point of the insertion or deletion to the end of the molecule.

To illustrate, assume that the DNA segment . . . CATCATCATCATCAT . . . is read in groups of three as . . . CAT-CAT-CAT-CAT-CAT. . . . If a nucleotide base – say, T – is inserted after the first C of the segment, the segment will then be read as . . . CTA-TCA-TCA-TCA-TCA. . . . From the point of the insertion onward, the sequence of encoded amino acids is altered. If, however, a total of three nucleotides is either added or deleted, the original reading frame will be maintained in the rest of the sequence. Additions or deletions of nucleotides in numbers other than three or multiples of three are called frame-shift mutations.

Gene mutations can occur spontaneously – that is, without being intentionally caused by humans. They can also be induced by ultraviolet light, X-rays, and other high-frequency electromagnetic radiation, as well as by exposure to certain mutagenic chemicals, such as mustard gas. The consequences of gene mutations may range from negligible to lethal. Mutations that change one or even several amino acids may have a small or undetectable effect on the organism's ability to survive and reproduce if the essential biological function of the coded protein is not hindered. But where an amino acid substitution affects the active site of an enzyme or in some other way modifies an essential function of a protein, the impact may be severe.

Newly arisen mutations are more likely to be harmful than beneficial to their carriers, because mutations are random events with respect to adaptation – that is, their occurrence is independent of any possible consequences. The allelic variants present in an existing population have already been subject to natural selection. They are present in the population because they improve the adaptation of their carriers, and their alternative alleles have been eliminated or kept at low frequencies by natural selection. A newly arisen mutant is likely to have been preceded by an identical mutation in the previous history of a population. If the previous mutant no longer exists in the population, it is a sign that the new mutant is not beneficial to the organism and is likely also to be eliminated.

This proposition can be illustrated with an analogy. Consider a sentence whose words have been chosen because together they express a certain idea. If single letters or words are replaced with others at random, most changes will be unlikely to improve the meaning of the sentence; very likely they will destroy it. The nucleotide sequence of a gene has been "edited" into its present form by natural selection because it "makes sense". If the sequence is changed at random, the "meaning" will rarely be improved and will often be hampered or destroyed.

Occasionally, however, a new mutation may increase the organism's adaptation. The probability of such an event happening is greater when organisms colonize a new territory or when environmental changes confront a population with new challenges. In these cases the established adaptation of a population is less than optimal, and there is greater opportunity for new mutations to be better adaptive. The consequences of mutations depend on the environment. Increased melanin pigmentation may be advantageous to inhabitants of tropical Africa, where dark skin protects them from the sun's ultraviolet radiation, but it is not beneficial at high latitudes, where the intensity of sunlight is low and pale skin facilitates the synthesis of vitamin D.

Mutation rates have been measured in a great variety of organisms, mostly for mutants that exhibit conspicuous effects. Mutation rates are generally lower in bacteria and other microorganisms than in more complex species. In humans and other multicellular organisms, the rate typically ranges from about 1 per 100,000 to 1 per 1,000,000 gametes. There is, however, considerable variation from gene to gene as well as from organism to organism.

Although mutation rates are low, new mutants appear continuously in nature, because there are many individuals in every species and many gene loci in every individual. The process of mutation provides each generation with many new genetic variations. Thus, it is not surprising that, when new environmental challenges arise, species are able to adapt to them. More than 200 insect and rodent species, for example, have developed resistance to the pesticide DDT in parts of the world where spraying has been intense. Although these animals had never before encountered this synthetic compound, they adapted to it rapidly by means of mutations that allowed them to survive in its presence. Similarly, many species of moths and butterflies in industrialized regions have shown an increase in the frequency of individuals with dark wings in response to environmental pollution, an adaptation known as industrial melanism (see below, Directional selection).

The resistance of disease-causing bacteria and parasites to antibiotics and other drugs is a consequence of the same process. When an individual receives an antibiotic that specifically kills the bacteria causing the disease – say, tuberculosis – the immense majority of the bacteria die, but one in a million may have a mutation that provides resistance to the antibiotic. These resistant bacteria will survive and multiply, and the antibiotic will no longer cure the disease. This is the reason that modern medicine treats bacterial diseases with cocktails of antibiotics. If the incidence of a mutation conferring resistance for a given antibiotic is one in a million, the incidence of one bacterium carrying three mutations, each conferring resistance to one of three antibiotics, is one in a trillion; such bacteria are far less likely to exist in any infected individual.

Chromosomal mutations

Chromosomes, which carry the DNA, are contained in the nucleus of each cell. Chromosomes come in pairs, with one member of each pair inherited from each parent. The two members of a pair are called homologous chromosomes. Each cell of an organism and all individuals of the same species have, as a rule, the same number of chromosomes. The reproductive cells (gametes) are an exception; they have only half as many chromosomes as the body (somatic) cells. But the number, size, and organization of chromosomes varies between species. The parasitic nematode *Parascaris univalens* has only 1 pair of chromosomes, whereas many species of butterflies have more than 100 pairs and some ferns more than 600. Even closely related organisms may vary considerably in the number of chromosomes. Species of spiny rats of the South American genus *Proechimys* range from 12 to 31 chromosome pairs.

Changes in the number, size, or organization of chromosomes within a species are termed chromosomal mutations, chromosomal abnormalities, or chromosomal aberrations. Changes in number may occur by the fusion of two chromosomes into one, by fission of one chromosome into two, or by addition or subtraction of one or more whole chromosomes or sets of chromosomes. (The condition in which an organism acquires one or more additional sets of chromosomes is called polyploidy.) Changes in the structure of chromosomes may occur by inversion, when a chromosomal segment rotates 180° within the same location; by duplication, when a segment is added; by deletion, when a segment is lost; or by translocation, when a segment changes from one location to another in the same or a different chromosome. These are the processes by which chromosomes evolve. Inversions, translocations, fusions, and fissions do not change the amount of DNA. The importance of these mutations in evolution is that they change the linkage relationships between genes. Genes that were closely linked to each other become separated and vice versa; this can affect their expression because genes are often transcribed sequentially, two or more at a time (see Chapter 4).

Dynamics of Genetic Change

Genetic Equilibrium: the Hardy–Weinberg Law

Genetic variation is present throughout natural populations of organisms. This variation is sorted out in new ways in each generation by the process of sexual reproduction, which recombines the chromosomes inherited from the two parents during the formation of the gametes that produce the following generation. But heredity by itself does not change gene frequencies. This principle is stated by the Hardy–Weinberg law, so called because it was independently discovered in 1908 by the British mathematician G.H. Hardy and the German physician Wilhelm Weinberg.

The Hardy–Weinberg law describes the genetic equilibrium in a population by means of an algebraic equation (Figure 5.2). It states that genotypes exist in certain frequencies that are a simple function of the allelic frequencies. The genotype equilibrium frequencies for any number of alleles are derived in the same way.

The genotype equilibrium frequencies are obtained by the Hardy–Weinberg law on the assumption that there is random mating – that is, the probability of a particular kind of mating is the same as the frequency of the genotypes of the two mating individuals. Random mating can occur with respect to most gene loci even though mates may be chosen according to particular characteristics. For example, people, choose their spouses according to all sorts of preferences concerning looks, personality, and the like. But, concerning the majority of genes, people's marriages are essentially random.

Assortative, or selective, mating takes place when the choice of mates is not random. Most marriages, for example, are assortative with respect to many social factors, so that members of any one social group tend to marry members of their

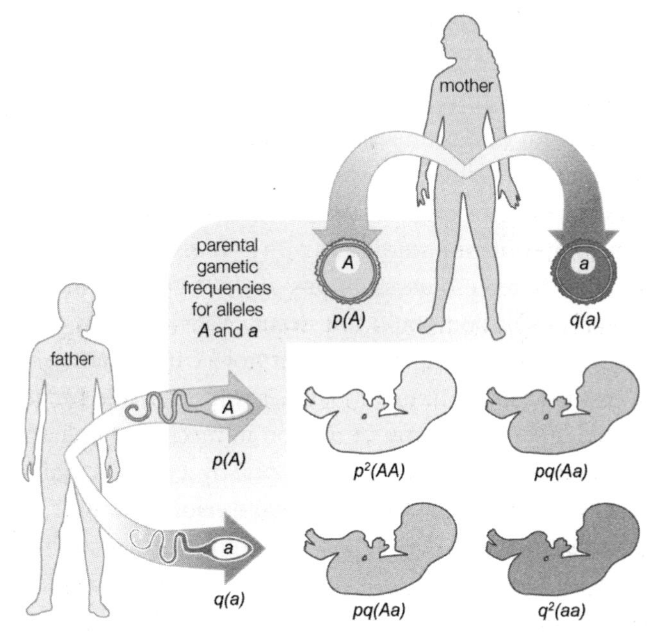

Figure 5.2 The Harvey–Weinberg law applied to two alleles. If there are two alleles, *A* and *a*, at a gene locus, three genotypes will be possible: *AA*, *Aa*, and *aa*. If the frequencies of the alleles are p and q, respectively, the equilibrium frequencies of the three genotypes will be given by $(p + q)2 = p^2 + 2pq + q^2$ for *AA*, *Aa*, and *aa*, respectively.

own group more often, and people from a different group less often, than would be expected from random mating. Consider the sensitive social issue of interracial marriage in a hypothetical community in which 80 per cent of the population is white and 20 per cent is black. With random mating, 32 per cent ($2 \times 0.80 \times 0.20 = 0.32$) of all marriages would be interracial, whereas only 4 per cent ($0.20 \times 0.20 = 0.04$) would be marriages between two blacks. These statistical expectations depart from typical observations even in modern society, as a result of persistent social customs that for evolutionists are examples of assortative mating. The most extreme form of assortative mating is self-fertilization, which occurs rarely in animals but is a common form of reproduction in many plant groups.

The Hardy–Weinberg law assumes that gene frequencies remain constant from generation to generation – that there is no gene mutation or natural selection and that populations are very large. But these assumptions are not correct; indeed, if they were, evolution could not occur. Why, then, is the law significant if its assumptions do not hold true in nature? The answer is that in evolutionary studies it plays a role similar to that of Newton's first law of motion in mechanics. Newton's first law says that a body not acted upon by a net external force remains at rest or maintains a constant velocity. In fact, there are always external forces acting upon physical objects, but the first law provides the starting point for the application of other laws. Similarly, organisms are subject to mutation, selection, and other processes that change gene frequencies, but the effects of these processes can be calculated by using the Hardy–Weinberg law as the starting point.

Changes in Gene Frequency

Mutation

Genetics has shown that mutation is the ultimate source of all hereditary variation. At the level of a single gene whose normal functional allele is *A*, it is known that mutation can change it to a non-functional recessive form, *a*. Such "forward mutation" is more frequent than "back mutation" (reversion), which converts *a* into *A*. Molecular analysis of specific examples of mutant recessive alleles has shown that they are generally a heterogeneous set of small structural changes in the DNA, located throughout the segment of DNA that constitutes that gene. Hence, in an example from medical genetics, the disease phenylketonuria is inherited as a recessive phenotype and is ascribed to a causative allele that generally can be called *k*. However, sequencing alleles of many independent cases of phenylketonuria has shown that this *k* allele is in fact a set of many different kinds of mutational changes, which can be in any of the protein-coding regions of that gene.

Recessive deleterious mutations are relatively rare, generally in the order of 1 per 105 or 106 mutant gametes per generation. Their constant occurrence over the generations, combined with the even greater rarity of back mutations, leads to a gradual accumulation in the population. This accumulation process is called mutational pressure.

Since mutational pressure to a deleterious recessive allele and selection pressure against the homozygous recessives are forces that act in opposite directions, another type of equilibrium, the mutation–selection equilibrium, is attained that effectively determines the frequency of an allele.

Gene flow

Gene flow, or gene migration, takes place when individuals migrate from one population to another and interbreed with its members. Gene frequencies are not changed for the species as a whole, but they change locally whenever different populations have different allele frequencies. In general, the greater the difference in allele frequencies between the resident and the migrant individuals, and the larger the number of migrants, the greater effect the migrants have in changing the genetic constitution of the resident population.

Genetic drift

Gene frequencies can change from one generation to another by a process of pure chance known as genetic drift. This occurs because the number of individuals in any population is finite, and thus the frequency of a gene may change in the following generation by accidents of sampling, just as it is possible to get more or fewer than 50 "heads" in 100 throws of a coin simply by chance.

The magnitude of the gene frequency changes due to genetic drift is inversely related to the size of the population – the larger the number of reproducing individuals, the smaller the effects of genetic drift. This inverse relationship between sample size and magnitude of sampling errors can be illustrated by referring again to tossing a coin. When a penny is tossed twice, 2 heads are not surprising. But it will be surprising, and suspicious, if 20 tosses all yield heads. The proportion of heads obtained in a series of throws approaches closer to 0.5 as the number of throws grows larger.

The relationship is the same in populations, although the important value here is not the actual number of individuals in the population but the "effective" population size. This is the number of individuals that produce offspring, because only reproducing individuals transmit their genes to the following generation. It is not unusual, in plants as well as animals, for some individuals to have large numbers of progeny while others have none. In seals, antelopes, baboons, and many other mammals, for example, a dominant male may keep a large harem of females at the expense of many other males who can find no mates. It often happens that the effective population size is substantially smaller than the number of individuals in any one generation.

The effects of genetic drift in changing gene frequencies from one generation to the next are quite small in most natural populations, which generally consist of thousands of reproducing individuals. The effects over many generations are more important. Indeed, in the absence of other processes of change (such as natural selection and mutation), populations would eventually become fixed, having one allele at each locus after the gradual elimination of all others. With genetic drift as the only force in operation, the probability of a given allele eventually reaching a frequency of 1 would be precisely the frequency of the allele – that is, an allele with a frequency of 0.8 would have an 80 per cent chance of ultimately becoming the only allele present in the population. The process would take a long time, however, because increases and decreases are likely to alternate with equal probability. More important, natural selection and other processes change gene frequencies in ways not governed by pure chance, so that no allele has an opportunity to become fixed as a consequence of genetic drift alone.

Genetic drift can have important evolutionary consequences when a new population becomes established by only a few individuals – a phenomenon known as the founder principle. Islands, lakes, and other isolated ecological sites are often colonized by one or very few seeds or animals of a species, which are transported there passively by wind, in the fur of larger animals, or in some other way. The allelic frequencies present in these few colonizers are likely to differ at many loci from those in the population they left, and those differences have a lasting impact on the evolution of the new population. The founder principle is one reason that species in neighbouring islands, such as those in the Hawaiian archipelago, are often more heterogeneous than species in comparable continental areas adjacent to one another.

Climatic or other conditions, if unfavourable, may on occasion drastically reduce the number of individuals in a population and even threaten it with extinction. Such occasional reductions are called population bottlenecks. The populations may later recover their typical size, but the allelic frequencies may have been considerably altered and thereby affect the future evolution of the species. Bottlenecks are more likely in relatively large animals and plants than in smaller ones, because populations of large organisms typically consist of fewer individuals. Primitive human populations of the past were subdivided into many small tribes that were time and again decimated by disease, war, and other disasters. Differences among current human populations in the allele frequencies of many genes – such as those determining the ABO and other blood groups – may have arisen at least in part as a consequence of bottlenecks in ancestral populations. Persistent population bottlenecks may reduce the overall genetic variation so greatly as to alter future evolution and endanger the survival of the species. A well-authenticated case is that of the cheetah, where no allelic variation whatsoever has been found among the many scores of gene loci studied.

Operation of Natural Selection in Populations

Natural selection as a process of genetic change

Natural selection refers to any reproductive bias favouring some genes or genotypes over others. Natural selection promotes the adaptation of organisms to the environments in which they live; any hereditary variant that improves the ability to survive and reproduce in an environment will increase in frequency over the generations, precisely because the organisms carrying such a variant will leave more descendants than those lacking it. Hereditary variants, favourable or not to the organisms, arise by mutation. Unfavourable ones are eventually eliminated by natural selection; their carriers leave no descendants, or leave fewer than those carrying alternative variants. Favourable mutations accumulate over the generations. The process continues indefinitely because the environments that organisms inhabit are forever changing.

Environments change physically – in their climate, configuration, and so on – but also biologically, because the predators, parasites, competitors, and food sources with which an organism interacts are themselves evolving.

Mutation, gene flow, and genetic drift are random processes with respect to adaptation; they change gene frequencies without regard for the consequences that such changes may have in the ability of the organisms to survive and reproduce. If these were the only processes of evolutionary change, the organization of living things would gradually disintegrate. The effects of such processes alone would be analogous to those of a mechanic who changed parts in an automobile engine at random, with no regard for the role of the parts in the engine. Natural selection keeps the disorganizing effects of mutation and other processes in check because it multiplies beneficial mutations and eliminates harmful ones.

Natural selection accounts not only for the preservation and improvement of the organization of living beings but also for their diversity. In different localities or in different circumstances, natural selection favours different traits – precisely those that make the organisms well adapted to their particular circumstances and ways of life.

The parameter used to measure the effects of natural selection is fitness (see above, The concept of natural selection), which can be expressed as an absolute or as a relative value. Consider a population consisting at a certain locus of three genotypes: A_1A_1, A_1A_2, and A_2A_2. Assume that on the average each A_1A_1 and each A_1A_2 individual produces one offspring but that each A_2A_2 individual produces two. One could use the average number of progeny left by each genotype as a measure of that genotype's absolute fitness and calculate the changes in gene frequency that would occur over the generations. (This, of course, requires knowing how many of the progeny survive to adulthood and reproduce.) Evolutionists, however, find it mathematically more convenient to use relative fitness values – which they represent by the letter w – in most calculations. They usually assign the value 1 to the genotype with the highest reproductive efficiency and calculate the other relative fitness values proportionally. For the example just used, the relative fitness of the A_2A_2 genotype would be $w = 1$ and that of each of the other two genotypes would be $w = 0.5$. A parameter related to fitness is the selection coefficient, often represented by the letter s, which is defined as $s = 1 - w$. The selection coefficient is a measure of the reduction in fitness of a genotype. The selection coefficients in the example are $s = 0$ for A_2A_2 and $s = 0.5$ for A_1A_1 and for A_1A_2.

The different ways in which natural selection affects gene frequencies are illustrated by the following examples.

Selection against one of the homozygotes

Suppose that one homozygous genotype, A_2A_2, has lower fitness than the other two genotypes, A_1A_1 and A_1A_2. This is the situation in many human diseases, such as phenylketonuria (PKU) and sickle cell anaemia, that are inherited in a recessive fashion and that require the presence of two deleterious mutant alleles for the trait to manifest. The heterozygotes and the homozygotes for the normal allele (A_1) have equal fitness, higher than that of the homozygotes for the deleterious mutant allele (A_2). The deleterious allele will continuously decrease in frequency until it has been eliminated. The rate of elimination is fastest when the selection coefficient is 1 (i.e. when the relative fitness is 0); this occurs with fatal diseases, such as untreated PKU, when the homozygotes die before the age of reproduction.

Because of new mutations, the elimination of a deleterious allele is never complete. A dynamic equilibrium frequency will exist when the number of new alleles produced by mutation is the same as the number eliminated by selection.

The mutation rate for many human recessive diseases is about 1 in 100,000. If the disease is fatal, the equilibrium frequency becomes about 1 recessive lethal mutant allele for every 300 normal alleles. That is roughly the frequency in human populations of alleles that in homozygous individuals, such as those with PKU, cause death before adulthood. The equilibrium frequency for a deleterious, but not lethal, recessive allele is much higher. Albinism, for example, is due to a recessive gene. The reproductive efficiency of people with albinism is, on average, about 0.9 that of normal individuals. Therefore, the selection coefficient becomes 1 in 100 genes rather than 1 in 300 as for a lethal allele.

For deleterious dominant alleles, if the gene is lethal even in single copy, all the genes are eliminated by selection in the same generation in which they arise, and the frequency of the gene in the population is the frequency with which it arises by mutation. One deleterious condition that is caused by a dominant allele present at low frequencies in human populations is achondroplasia, the most common cause of dwarfism. People with achondroplasia reproduce only 20 per cent as efficiently as normal individuals.

Overdominance

In many instances heterozygotes have a higher degree of fitness than homozygotes for one or the other allele. This situation, known as heterosis or overdominance, leads to the stable coexistence of both alleles in the population and hence contributes to the widespread genetic variation found in populations of most organisms.

A particularly interesting example of heterozygote superiority among humans is provided by the gene responsible for sickle cell anaemia. In this condition, a mutant allele, Hb^S modifies the properties of haemoglobin so that homozygotes (Hb^SHb^S), suffer from a severe form of anaemia that in most cases leads to death before the age of reproduction.

The Hb^S allele occurs with a high frequency in some African and Asian populations. This was formerly puzzling because the severity of the anaemia, representing a strong natural selection against homozygotes, should have eliminated the defective allele. But researchers noticed that the Hb^S allele occurred at high frequency precisely in regions of the world where a particularly severe form of malaria, which is caused by the parasite *Plasmodium falciparum*, was endemic. It was hypothesized that the heterozygotes, who had both the normal and mutant alleles (Hb^AHb^S), were resistant to malaria, whereas the homozygotes with normal alleles (Hb^AHb^A) were not. In malaria-infested regions the heterozygotes survived better than either of the homozygotes, which were more likely to die from either malaria (Hb^AHb^A homozygotes) or anaemia (Hb^SHb^S homozygotes). This hypothesis has been confirmed in various ways. Most significant is that most hospital patients with severe or fatal forms of malaria are homozygotes Hb^AHb^A. In a study of 100 children who died from malaria, only 1 was found to be a heterozygote, whereas 22 were expected to be so, according to the frequency of the Hb^S allele in the population.

Frequency-dependent selection

The fitness of genotypes can change when the environmental conditions change. White fur may be protective for a bear living on the Arctic snows but not for one living in a Russian forest; there, an allele coding for brown pigmentation may be favoured over one that codes for white. The environment of an organism includes not only the climate and other physical features but also the organisms of the same or different species with which it is associated.

Changes in genotypic fitness are associated with the density of the organisms present. Insects and other short-lived organisms experience enormous yearly oscillations in density. Some genotypes may possess high fitness in the spring, when the population is rapidly expanding, because such genotypes yield more prolific individuals. Other genotypes may be favoured during the summer, when populations are dense, because these genotypes make for better competitors, ones more successful at securing limited food resources. Still others may be at an advantage during the winter months, because they increase the population's hardiness, or ability to withstand the inclement conditions that kill most members of the other genotypes.

The fitness of genotypes can also vary according to their relative numbers, and genotype frequencies may change as a consequence. This is known as frequency-dependent selection. Particularly interesting is the situation in which genotypic fitnesses are inversely related to their frequencies. Assume that two genotypes, *A* and *B*, have fitnesses related to their frequencies in such a way that the fitness of either genotype increases when its frequency decreases and vice versa. When *A* is rare, its fitness is high, and therefore *A* increases in frequency. As it becomes more and more common, however, the fitness of *A* gradually decreases, so that its increase in frequency eventually comes to a halt. A stable polymorphism occurs at the frequency where the two genotypes, *A* and *B*, have identical fitnesses.

In natural populations of animals and plants, frequency-dependent selection is very common and may contribute importantly to the maintenance of genetic polymorphism.

Frequency-dependent selection may arise because the environment is heterogeneous and because different genotypes can better exploit different sub-environments. When a genotype is rare, the sub-environments that it exploits better will be relatively abundant. But as the genotype becomes common, its favoured sub-environment becomes saturated. That genotype must then compete for resources in sub-environments that are optimal for other genotypes.

It follows then that a mixture of genotypes exploits the environmental resources better than a single genotype. Plant breeders know that mixed plantings (a mixture of different strains) are more productive than single stands (plantings of one strain only), although farmers avoid them for reasons such as increased harvesting costs.

Sexual preferences can also lead to frequency-dependent selection. It has been demonstrated in some insects, birds, mammals, and other organisms that the mates preferred are precisely those that are rare. People also appear to experience this rare-mate advantage – fair-haired men may seem attractively exotic to brunettes, or dark-haired men to blondes.

Types of selection

Non-random mating

Many species engage in alternatives to random mating as normal parts of their cycle of sexual reproduction. An important exception is sexual selection, in which an individual chooses a mate on the basis of some aspect of the mate's phenotype. The selection can be based on some display feature such as bright feathers, or it may be a simple preference for a phenotype identical to the individual's own (positive assortative mating).

Two other important exceptions are inbreeding (mating with relatives) and enforced outbreeding. Both can shift the equilibrium proportions expected under Hardy–Weinberg calculations. For example, inbreeding increases the proportions of homozygotes, and the most extreme form of inbreeding, self-fertilization, eventually eliminates all heterozygotes.

Inbreeding and outbreeding are evolutionary strategies adopted by plants and animals living under certain conditions. Outbreeding brings gametes of different genotypes together, and the resulting individual differs from the parents. Increased levels of variation provide more evolutionary flexibility. All the showy colours and shapes of flowers are to promote this kind of exchange. In contrast, inbreeding maintains uniform genotypes, a strategy successful in stable ecological habitats.

In humans, various degrees of inbreeding have been practised in different cultures. In most cultures today, matings of first cousins are the closest form of inbreeding condoned by society. Apart from ethical considerations, a negative outcome of inbreeding is that it increases the likelihood of homozygosity of deleterious recessive alleles originating from common ancestors, called homozygosity by descent. The inbreeding coefficient F is a measure of the likelihood of homozygosity by descent; for example, in first-cousin marriages, $F = 1/16$. A large proportion of recessive hereditary diseases can be traced to first-cousin marriages and other types of inbreeding.

Stabilizing selection

Natural selection can be studied by analysing its effects on changing gene frequencies, but it can also be explored by examining its effects on the observable characteristics – or phenotypes – of individuals in a population. Distribution scales of phenotypic traits such as height, weight, number of progeny, or longevity typically show greater numbers of individuals with intermediate values and fewer and fewer toward the extremes – this is the so-called normal distribution. When individuals with intermediate phenotypes are favoured and extreme phenotypes are selected against, the selection is said to be stabilizing. The range and distribution of phenotypes then remain approximately the same from one generation to another. Stabilizing selection is very common. The individuals that survive and reproduce more successfully are those that have intermediate phenotypic values. Mortality among newborn infants, for example, is highest when they are either very small or very large; infants of intermediate size have a greater chance of surviving.

Stabilizing selection is often noticeable after artificial selection. Breeders choose hens that produce larger eggs, cows that yield more milk, and corn with higher protein content. But the selection must be continued or reinstated from time to time, even after the desired goals have been achieved. If it is stopped altogether, natural selection gradually takes effect and turns the traits back toward their original intermediate value.

As a result of stabilizing selection, populations often maintain a steady genetic constitution with respect to many traits. This attribute of populations is called genetic homeostasis.

Directional selection

The distribution of phenotypes in a population sometimes changes systematically in a particular direction. The physical and biological aspects of the environment are continuously changing, and over long periods of time the changes may be substantial. The climate and even the configuration of the land or waters vary incessantly. Changes also take place in the biotic conditions – that is, in the other organisms present, whether predators, prey, parasites, or competitors. Genetic changes occur as a consequence, because the genotypic fitnesses may shift so that different sets of alleles are favoured. The opportunity for directional selection also arises when organisms colonize new environments where the conditions are different from those of their original habitat. In addition, the appearance of a new favourable allele or a new genetic combination may prompt directional changes as the new genetic constitution replaces the pre-existing one.

The process of directional selection takes place in spurts. The replacement of one genetic constitution with another changes the genotypic fitnesses at other loci, which then change in their allelic frequencies, thereby stimulating additional changes, and so on in a cascade of consequences.

Directional selection is possible only if there is genetic variation with respect to the phenotypic traits under selection. Natural populations contain large stores of genetic variation, and these are continuously replenished by additional new variants that arise by mutation. The nearly universal success of artificial selection and the rapid response of natural populations to new environmental challenges are evidence that existing variation provides the necessary materials for directional selection.

In modern times human actions have been an important stimulus to this type of selection. Human activity transforms the environments of many organisms, which rapidly respond to the new environmental challenges through directional selection. Well-known instances are the many cases of insect resistance to pesticides, which are synthetic substances not present in the natural environment. When a new insecticide is first applied to control a pest, the results are encouraging because a small amount of the insecticide is sufficient to bring the pest organism under control. As time passes, however, the amount required to achieve a certain level of control must be increased again and again until finally it becomes ineffective or economically impractical. This occurs because organisms become resistant to the pesticide through directional selection. The resistance of the housefly to DDT was first reported in 1947. Resistance to one or more pesticides has since been recorded in several hundred species of insects and mites.

Another example is the phenomenon of industrial melanism (mentioned above in the section Gene mutations), which is exemplified by the gradual darkening of the wings of many species of moths and butterflies living in woodlands darkened by industrial pollution. The best-investigated case is the peppered moth in England. Until the middle of the 19th century, these moths were uniformly peppered light grey. Darkly pigmented variants were detected first in 1848 in Manchester and shortly afterwards in other industrial regions where the vegetation was blackened by soot and other pollutants. By the middle of the 20th century, the dark varieties had almost completely replaced the lightly pigmented forms in many polluted areas, while in unpolluted regions light moths continued to be the most common. The shift from light to dark moths was an example of directional selection brought about by bird predators. On lichen-covered tree trunks, the light-grey moths are well camouflaged, whereas the dark ones are conspicuously visible and therefore fall victim to the birds. The opposite is the case on trees darkened by pollution.

Over geologic time, directional selection leads to major changes in morphology and ways of life. Evolutionary changes that persist in a more or less continuous fashion over long periods of time are known as evolutionary trends. Directional evolutionary changes increased the cranial capacity of the human lineage from the small brain of *Australopithecus* – human ancestors of 3 million years ago – which was less than 500 cm^3 in volume, to a brain nearly three times as large in modern humans. The evolution of the horse from more than 50 million years ago to modern times is another well-studied example of directional selection.

Diversifying selection

Two or more divergent phenotypes in an environment may be favoured simultaneously by diversifying selection. No natural environment is homogeneous; rather, the environment of any plant or animal population is a mosaic consisting of more or less dissimilar sub-environments. There is heterogeneity with respect to climate, food resources, and living space. Also, the heterogeneity may be temporal, with change occurring over time, as well as spatial. Species cope with environmental heterogeneity in diverse ways. One strategy is genetic monomorphism, the selection of a generalist genotype that is well adapted to all the sub-environments encountered by the species. Another strategy is genetic polymorphism, the selection of a diversified gene pool that yields different genotypes, each adapted to a specific sub-environment.

There is no single plan that prevails in nature. Sometimes the most efficient strategy is genetic monomorphism to confront temporal heterogeneity but polymorphism to confront spatial heterogeneity. If the environment changes in time, or if it is unstable relative to the lifespan of the organisms, each individual will have to face diverse environments appearing one after the other. A series of genotypes, each well adapted to one or another of the conditions that prevail at various times, will not succeed very well, because each organism will fare well at one period of its life but not at others. A better strategy is to have a population with one or a few genotypes that survive well in all the successive environments.

If the environment changes from place to place, the situation is likely to be different. Although a single genotype, well adapted to the various environmental patches is a possible strategy, a variety of genotypes, with some individuals optimally adapted to each sub-environment, might fare still better. The ability of the population to exploit the environmental patchiness is thereby increased. Diversifying selection refers to the situation in which natural selection favours different genotypes in different sub-environments.

The efficiency of diversifying natural selection is quite apparent in circumstances in which populations living a short distance apart have become genetically differentiated. In one example, populations of bent grass (*Agrostis*) can be found growing on heaps of mining refuse heavily contaminated with heavy metals such as lead and copper. The soil has become so contaminated that it is toxic to most plants, but the dense stands of bent grass growing over these refuse heaps have been shown to possess genes that make them resistant to high concentrations of lead and copper. But only a few metres from the contaminated soil can be found bent grass plants that are not resistant to these metals. Bent grasses reproduce primarily by cross-pollination, so that the resistant grass receives wind-borne pollen from the neighbouring non-resistant plants. Yet they maintain their genetic differentiation because non-resistant seedlings are unable to grow in the contaminated soil and, in nearby uncontaminated soil, the non-resistant seedlings outgrow the resistant ones. The evolution of these resistant strains has taken place since the mines were first opened, a period of less than 400 years.

Protective morphologies and protective colouration exist in many animals as a defence against predators or as a cover against prey. Sometimes an organism mimics the appearance of a different one for protection. Diversifying selection often occurs in association with mimicry. A species of swallowtail butterfly, *Papilio dardanus*, is endemic in tropical and southern Africa. Males have yellow and black wings, with characteristic tails in the second pair of wings. But females in many localities are conspicuously different from males; their wings lack tails and have colour patterns that vary from place to place. The explanation for these differences stems from the fact that *P. dardanus* can be eaten safely by birds. Many other butterfly species are noxious to birds, and so they are carefully avoided as food. In localities where *P. dardanus* coexists with noxious butterfly species, the *P. dardanus* females have evolved an appearance that mimics the noxious species. Birds confuse the mimics with their models and do not prey on them. In different localities the females mimic different species; in some areas two or even three different female forms exist, each mimicking different noxious species. Diversifying selection has resulted in different phenotypes of *P. dardanus* as a protection from bird predators.

Sexual selection

Mutual attraction between the sexes is an important factor in reproduction. The males and females of many animal species are similar in size and shape except for the sexual organs and secondary sexual characteristics such as the breasts of female mammals. In some species, however, the sexes exhibit striking dimorphism. Particularly in birds and mammals, the males are often larger and stronger, more brightly coloured, or endowed with conspicuous adornments. But bright colours make animals more visible to predators – the long plumage of male peacocks and birds of paradise and the enormous antlers of aged male deer are cumbersome loads in the best of cases. Darwin knew that natural selection could not be expected to favour the evolution of disadvantageous traits, and he was able to offer a solution to this problem. He proposed that such traits arise by "sexual selection", which "depends not on a struggle for existence in relation to other organic beings or to external conditions but on a struggle between the individuals of one sex, generally the males, for the possession of the other sex".

The concept of sexual selection as a special form of natural selection is easily explained. Other things being equal, organisms more proficient in securing mates have higher fitness.

There are two general circumstances leading to sexual selection. One is the preference shown by one sex (often the females) for individuals of the other sex that exhibit certain traits. The other is increased strength (usually among the males) that yields greater success in securing mates.

The presence of a particular trait among the members of one sex can make them somehow more attractive to the opposite sex. This type of "sex appeal" has been experimentally demonstrated in all sorts of animals, from vinegar flies to pigeons, mice, dogs, and rhesus monkeys. When, for example, *Drosophila* flies, some with yellow bodies as a result of spontaneous mutation and others with the normal yellowish grey pigmentation, are placed together, normal males are preferred over yellow males by females with either body colour.

Sexual selection can also come about because a trait – the antlers of a stag, for example – increases prowess in competition with members of the same sex. Stags, rams, and bulls use antlers or horns in contests of strength; a winning male usually secures more female mates. Therefore, sexual selection may lead to increased size and aggressiveness in males. Male baboons are more than twice as large as females, and the behaviour of the docile females contrasts with that of the aggressive males. A similar dimorphism occurs in the northern sea lion, where males weigh about 1,000 kg (2,200 pounds), about three times as much as females. The males fight fiercely in their competition for females; large, battle-scarred males occupy their own rocky islets, each holding a harem of as many as 20 females. Among many mammals that live in packs, troops, or herds – such as wolves, horses, and buffaloes – there usually is a hierarchy of dominance based on age and strength, with males that rank high in the hierarchy doing most of the mating.

Kin selection and reciprocal altruism

The apparent altruistic behaviour of many animals is, like some manifestations of sexual selection, a trait that at first seems incompatible with the theory of natural selection. Altruism is a form of behaviour that benefits other individuals at the expense of the one that performs the action; the fitness of the altruist is diminished by its behaviour, whereas individuals that act selfishly benefit from it at no cost to themselves. Accordingly, it might be expected that natural selection would foster the development of selfish behaviour and eliminate altruism. This conclusion is not so compelling when it is noticed that the beneficiaries of altruistic behaviour are usually relatives. They all carry the same genes, including the genes that promote altruistic behaviour. Altruism may evolve by kin selection, which is simply a type of natural selection in which relatives are taken into consideration when evaluating an individual's fitness.

Natural selection favours genes that increase the reproductive success of their carriers, but it is not necessary that all individuals that share a given genotype have higher reproductive success. It suffices that carriers of the genotype reproduce more successfully on the average than those possessing alternative genotypes. A parent shares half of its genes with each progeny, so a gene that promotes parental altruism is favoured by selection if the behaviour's cost to the parent is less than half of its average benefits to the progeny. Such a gene will be more likely to increase in frequency through the generations than an alternative gene that does not promote altruistic behaviour. Parental care is, therefore, a form of altruism readily explained by kin selection. The parent spends some energy caring for the progeny because it increases the reproductive success of the parent's genes.

Kin selection extends beyond the relationship between parents and their offspring. It facilitates the development of altruistic behaviour when the energy invested, or the risk incurred, by an individual is compensated in excess by the benefits ensuing to relatives. The closer the relationship between the beneficiaries and the altruist and the greater the number of beneficiaries, the higher the risks and efforts warranted in the altruist. Individuals that live together in a herd or troop usually are related and often behave toward each other in this way. Adult zebras, for instance, will turn toward an attacking predator to protect the young in the herd rather than fleeing to protect themselves.

Altruism also occurs among unrelated individuals when the behaviour is reciprocal and the altruist's costs are smaller than the benefits to the recipient. This reciprocal altruism is found in the mutual grooming of chimpanzees and other primates as they clean each other of lice and other pests. Another example appears in flocks of birds that post sentinels to warn of danger. A crow sitting in a tree watching for predators while the rest of the flock forages incurs a small loss by not feeding, but this loss is well compensated by the protection it receives when it itself forages and others of the flock stand guard.

A particularly valuable contribution of the theory of kin selection is its explanation of the evolution of social behaviour among ants, bees, wasps, and other social insects. In honeybee populations, for example, the female workers build the hive, care for the young, and gather food, but they are sterile; the queen bee alone produces progeny. It would seem that the workers' behaviour would in no way be promoted or maintained by natural selection. Any genes causing such behaviour would seem likely to be eliminated from the population, because individuals exhibiting the behaviour increase not their own reproductive success but that of the queen. The situation is, however, more complex.

Queen bees produce some eggs that remain unfertilized and develop into males, or drones, having a mother but no father.

Their main role is to engage in the nuptial flight during which one of them fertilizes a new queen. Other eggs laid by queen bees are fertilized and develop into females, the large majority of which are workers. A queen typically mates with a single male once during her lifetime; the male's sperm is stored in the queen's spermatheca, from which it is gradually released as she lays fertilized eggs. All the queen's female progeny therefore have the same father, so that workers are more closely related to one another and to any new sister queen than they are to the mother queen. The female workers receive one-half of their genes from the mother and one-half from the father, but they share among themselves three-quarters of their genes. The half of the set from the father is the same in every worker, because the father had only one set of genes rather than two to pass on (the male developed from an unfertilized egg, so all his sperm carry the same set of genes). The other half of the workers' genes come from the mother, and on the average half of them are identical in any two sisters. Consequently, with three-quarters of her genes present in her sisters but only half of her genes able to be passed on to a daughter, a worker's genes are transmitted one and a half times more effectively when she raises a sister (whether another worker or a new queen) than if she produces a daughter of her own.

Species and Speciation

Concept of Species

Darwin sought to explain the splendid multiformity of the living world – thousands of organisms of the most diverse kinds, from lowly worms to spectacular birds of paradise, from yeasts and moulds to oaks and orchids. His *On the Origin of Species by Means of Natural Selection* (1859) is a sustained argument showing that the diversity of organisms and their characteristics can be explained as the result of natural processes.

Species come about as the result of gradual change prompted by natural selection. Environments are continually changing in time, and they differ from place to place. Natural selection therefore favours different characteristics in different situations. The accumulation of differences eventually yields different species.

Everyday experience teaches that there are different kinds of organisms and also teaches how to identify them. Everyone knows that people belong to the human species and are different from cats and dogs, which in turn are different from each other. There are differences between people, as well as between cats and dogs, but individuals of the same species are considerably more similar among themselves than they are to individuals of other species.

External similarity is the common basis for identifying individuals as being members of the same species. Nevertheless, there is more to a species than outward appearance. A bulldog, a terrier, and a golden retriever are very different in appearance, but they are all dogs because they can interbreed. People can also interbreed with one another, and so can cats with other cats, but people cannot interbreed with dogs or cats, nor can these with each other. It is clear then that, although species are usually identified by appearance, there is something basic, of great biological significance, behind similarity of appearance – individuals of a species are able to interbreed with one another but not with members of other species. This is expressed in the following definition: Species are groups of interbreeding natural populations that are reproductively isolated from other such groups. (For an explanation and discussion of this concept, see below Reproductive isolation.)

The ability to interbreed is of great evolutionary importance, because it determines that species are independent evolutionary units. Genetic changes originate in single individuals; they can spread by natural selection to all members of the species but not to individuals of other species. Individuals of a species share a common gene pool that is not shared by individuals of other species. Different species have independently evolving gene pools because they are reproductively isolated.

Although the criterion for deciding whether individuals belong to the same species is clear, there may be ambiguity in practice for two reasons. One is lack of knowledge – it may not be known for certain whether individuals living in different sites belong to the same species, because it is not known whether they can naturally interbreed. The other reason for ambiguity is rooted in the nature of evolution as a gradual process. Two geographically separate populations that at one time were members of the same species later may have diverged into two different species. Since the process is gradual, there is no particular point at which it is possible to say that the two populations have become two different species.

A related situation pertains to organisms living at different times. There is no way to test if today's humans could interbreed with those who lived thousands of years ago. It seems reasonable that living people, or living cats, would be able to interbreed with people, or cats, exactly like those that lived a few generations earlier. But what about ancestors removed by a thousand or a million generations? The ancestors of modern humans that lived 500,000 years ago (about 20,000 generations) are classified as the species *Homo erectus*. There is no exact time at which *H. erectus* became *H. sapiens*, but it would not be appropriate to classify remote human ancestors and modern humans in the same species just because the changes from one generation to the next were small. It is useful to distinguish between the two groups by means of different species names, just as it is useful to give different names to childhood and adulthood even though no single moment can separate one from the other. Biologists distinguish species in organisms that lived at different times by means of a commonsense morphological criterion: If two organisms differ from each other in form and structure about as much as do two living individuals belonging to two different species, they are classified in separate species and given different names.

The definition of species given above applies only to organisms able to interbreed. Bacteria and cyanobacteria (blue-green algae), for example, reproduce not sexually but by fission. Organisms that lack sexual reproduction are classified into different species according to criteria such as external morphology, chemical and physiological properties, and genetic constitution.

Origin of Species

Reproductive isolation

Among sexual organisms, individuals that are able to interbreed belong to the same species. The biological properties of organisms that prevent interbreeding are called reproductive isolating mechanisms (RIMs). Oaks on different islands, minnows in different rivers, or squirrels in different mountain ranges cannot interbreed because they are physically separated, not necessarily because they are biologically incompatible. Geographic separation, therefore, is not a RIM.

There are two general categories of RIMs: prezygotic, or those that take effect before fertilization, and postzygotic, those that take effect afterwards. Prezygotic RIMs prevent the formation of hybrids between members of different populations through ecological, temporal, ethological (behavioral), mechanical, and gametic isolation. Postzygotic RIMs reduce the viability or fertility of hybrids or their progeny.

Ecological isolation

Populations may occupy the same territory but live in different habitats and so not meet. The *Anopheles maculipennis* group consists of six mosquito species, some of which are involved in the transmission of malaria. Although the species are virtually indistinguishable morphologically, they are isolated reproductively, in part because they breed in different habitats. Some breed in brackish water, others in running fresh water, and still others in stagnant fresh water.

Temporal isolation

Populations may mate or flower at different seasons or different times of day. Three tropical orchid species of the genus *Dendrobium* each flower for a single day; the flowers open at dawn and wither by nightfall. Flowering occurs in response to certain meteorological stimuli, such as a sudden storm on a hot day. The same stimulus acts on all three species, but the lapse between the stimulus and flowering is 8 days in one species, 9 in another, and 10 or 11 in the third. Interspecific fertilization is impossible because, at the time the flowers of one species open, those of the other species have already withered or have not yet matured.

A peculiar form of temporal isolation exists between pairs of closely related species of cicadas, in which one species of each pair emerges every 13 years, the other every 17 years. The two species of a pair may be sympatric (live in the same territory), but they have an opportunity to form hybrids only once every 221 (or 13 × 17) years.

Ethological (behavioral) isolation

Sexual attraction between males and females of a given species may be weak or absent. In most animal species, members of the two sexes must first search for each other and come together. Complex courtship rituals then take place, with the male often taking the initiative and the female responding. This in turn generates additional actions by the male and responses by the female, and eventually there is copulation, or sexual intercourse (or, in the case of some aquatic organisms, release of the sex cells for fertilization in the water). These elaborate rituals are specific to a species and play a significant part in species recognition. If the sequence of events in the search–courting–mating process is rendered disharmonious by either of the two sexes, then the entire process will be interrupted. Courtship and mating rituals have been extensively analysed in some mammals, birds, and fishes and in a number of insect species.

Ethological isolation is often the most potent RIM to keep animal species from interbreeding. It can be remarkably strong even among closely related species. The vinegar flies *Drosophila serrata*, *D. birchii*, and *D. dominicana* are three sibling species (that is, species nearly indistinguishable morphologically) that are endemic in Australia and on the islands of New Guinea and New Britain. In many areas these three species occupy the same territory, but no hybrids are known to occur in nature. The strength of their ethological isolation has been tested in the laboratory by placing together groups of females and males in various combinations for several days. When the flies were all of the same species but the female and male groups each came from different geographic origins, a large majority of the females (usually 90 per cent or more) were fertilized. But no inseminations or very few (less than 4 per cent) took place when males and females were of different species, whether from the same or different geographic origins.

It should be added that the rare interspecific inseminations that did occur among the vinegar flies produced hybrid adult individuals in very few instances, and the hybrids were always sterile. This illustrates a common pattern – reproductive isolation between species is maintained by several RIMs in succession; if one breaks down, others are still present. In addition to ethological isolation, failure of the hybrids to survive and hybrid sterility (see below Hybrid inviability and Hybrid sterility) prevent successful breeding between members of the three *Drosophila* species and between many other animal species as well.

Species recognition during courtship involves stimuli that may be chemical (olfactory), visual, auditory, or tactile. Pheromones are specific substances that play a critical role in recognition between members of a species; they have been chemically identified in such insects as ants, moths, butterflies, and beetles and in such vertebrates as fish, reptiles, and mammals. The "songs" of birds, frogs, and insects (the last of which produce these sounds by vibrating or rubbing their wings) are species recognition signals. Some form of physical contact or touching occurs in many mammals but also in *Drosophila* and other insects.

Mechanical isolation

Copulation is often impossible between different animal species because of the incompatible shape and size of the genitalia. In plants, variations in flower structure may impede pollination. Two species of sage from California provide an example. The two-lipped flowers of *Salvia mellifera* have stamens and style (respectively, the male structure that produces the pollen and the female structure that bears the pollen-receptive surface, the stigma) in the upper lip, whereas *S. apiana* has long stamens and style and a specialized floral configuration. *S. mellifera* is pollinated by small or medium-sized bees that carry pollen on their backs from flower to flower. *S. apiana*, however, is pollinated by large carpenter bees and bumblebees that carry the pollen on their wings and other body parts. Even if the pollinators of one species visit flowers of the other, pollination cannot occur because the pollen does not come into contact with the style of the alternative species.

Gametic isolation

Marine animals often discharge their eggs and sperm into the surrounding water, where fertilization takes place. Gametes of different species may fail to attract one another. For example, the sea urchins *Strongylocentrotus purpuratus* and *S. franciscanus* can be induced to release their eggs and sperm simultaneously, but most of the fertilizations that result are between eggs and sperm of the same species. In animals with internal fertilization, sperm cells may be unable to function in the sexual ducts of females of different species. In plants, pollen grains of one species typically fail to germinate on the stigma of another species, so that the pollen tubes never reach the ovary where fertilization would occur.

Hybrid inviability

Occasionally, prezygotic mechanisms are absent or break down so that interspecific zygotes (fertilized eggs) are formed. These zygotes, however, often fail to develop into mature individuals. The hybrid embryos of sheep and goats, for example, die in the early developmental stages before birth. Hybrid inviability is common in plants, whose hybrid seeds often fail to germinate or die shortly after germination.

Hybrid sterility

Hybrid zygotes sometimes develop into adults, such as mules (hybrids between female horses and male donkeys), but the adults fail to develop functional gametes and are sterile.

Hybrid breakdown

In plants more than in animals, hybrids between closely related species are sometimes partially fertile. Gene exchange may nevertheless be inhibited because the offspring are poorly viable or sterile. Hybrids between the cotton species *Gossypium barbadense*, *G. hirsutum*, and *G. tomentosum* appear vigorous and fertile, but their progenies die in seed or early in development, or they develop into sparse, weak plants.

A model of speciation

Because species are groups of populations reproductively isolated from one another, asking about the origin of species is equivalent to asking how reproductive isolation arises between populations. Two theories have been advanced to answer this question. One theory considers isolation as an accidental by-product of genetic divergence. Populations that become genetically less and less alike (as a consequence, for example, of adaptation to different environments) may eventually be unable to interbreed because their gene pools are disharmonious. The other theory regards isolation as a product of natural selection. Whenever hybrid individuals are less fit than non-hybrids, natural selection will directly promote the development of RIMs. This occurs because genetic variants interfering with hybridization have greater fitness than those favouring hybridization, given that the latter are often present in hybrids with poor fitness.

These two theories of the origin of reproductive isolation are not mutually exclusive. Reproductive isolation may indeed come about incidentally to genetic divergence between separated populations. Consider, for example, the evolution of many endemic species of plants and animals in the Hawaiian archipelago. The ancestors of these species arrived on these islands several million years ago. There they evolved as they became adapted to the environmental conditions and colonizing opportunities present. Reproductive isolation between the populations evolving in Hawaii and the populations on continents was never directly promoted by natural selection because their geographic remoteness forestalled any opportunities for hybridizing. Nevertheless, reproductive isolation became complete in many cases as a result of gradual genetic divergence over thousands of generations.

Frequently, however, the course of speciation involves the processes postulated by both theories – reproductive isolation starts as a by-product of gradual evolutionary divergence but is completed by natural selection directly promoting the evolution of prezygotic RIMs.

The separate sets of processes identified by the two speciation theories may be seen, therefore, as different stages in the splitting of an evolutionary lineage into two species. The splitting starts when gene flow is somehow interrupted between two populations. It is necessary that gene flow be interrupted, because otherwise the two groups of individuals would still share in a common gene pool and fail to become genetically different. Interruption may be due to geographic separation, or it may be initiated by some genetic change that affects some individuals of the species but not others living in the same territory. The two genetically isolated groups are likely to become more and more different as time goes on. Eventually, some incipient reproductive isolation may take effect because the two gene pools are no longer adapting in concert. Hybrid individuals, which carry genes combined from the two gene pools, will therefore experience reduced viability or fertility.

The circumstances just described may persist for so long that the populations become completely differentiated into separate species. In both animals and plants, however, it happens quite commonly that opportunities for hybridization arise between two populations that are becoming genetically differentiated. Two outcomes are possible. One is that the hybrids manifest little or no reduction of fitness, so that gene exchange between the two populations proceeds freely, eventually leading to their integration into a single gene pool. The second possible outcome is that reduction of fitness in the hybrids is sufficiently large for natural selection to favour the emergence of prezygotic RIMs, preventing the formation of hybrids altogether. This situation may be identified as the second stage in the speciation process.

How natural selection brings about the evolution of prezygotic RIMs can be understood in the following way. Beginning with two populations, P1 and P2, assume that there are gene variants in P1 that increase the probability that P1 individuals will choose P1 rather than P2 mates. Such gene variants will increase in frequency in the P1 population, because they are more often present in the progenies of P1 × P1 matings, which have normal fitness. The alternative genetic variants that do not favour P1 × P1 matings will be more often present in the progenies of P1 × P2 matings, which have lower fitness. The same process will enhance the frequency in the P2 population of genetic variants that lead P2 individuals to choose P2 rather than P1 mates. Prezygotic RIMs may therefore evolve in both populations and lead to their becoming two separate species.

The two stages of the process of speciation can be characterized, finally, by outlining their distinctions. The first stage primarily involves the appearance of postzygotic RIMs as accidental by-products of overall genetic differentiation rather than as express targets of natural selection. The second stage involves the evolution of prezygotic RIMs that are directly promoted by natural selection. The first stage may come about suddenly, in one or a few generations, rather than as a long, gradual process. The second stage follows the first in time but need not always be present.

Geographic speciation

One common mode of speciation is known as geographic, or allopatric (in separate territories), speciation. The general model of the speciation process advanced in the previous section applies well to geographic speciation. The first stage begins as a result of geographic separation between populations. This may occur when a few colonizers reach a geographically separate habitat, perhaps an island, lake, river, isolated valley, or mountain range. Alternately, a population may be split into two geographically separate ones by topographic changes, such as the disappearance of a water connection between two lakes, or by an invasion of competitors, parasites, or predators into the intermediate zone. If these types of geographic separation continue for some time, postzygotic RIMs may appear as a result of gradual genetic divergence.

In the second stage, an opportunity for interbreeding may later be brought about by topographic changes re-establishing continuity between the previously isolated territories or by ecological changes once again making the intermediate territory habitable for the organisms. If postzygotic RIMs that evolved during the separation period sufficiently reduce the fitness of hybrids of the two populations, natural selection will foster the development of prezygotic RIMs, and the two populations may go on to evolve into two species despite their occupying the same geographic territory.

Investigation has been made of many populations that are in the first stage of geographic speciation. There are fewer well-documented instances of the second stage, presumably because this occurs fairly rapidly in evolutionary time.

Both stages of speciation are present in a group of six closely related species of New World *Drosophila* flies that have been extensively studied by evolutionists for several decades. Two of these sibling species, *D. willistoni* and *D. equinoxialis*, each consist of groups of populations in the first stage of speciation and are identified as different subspecies. Two *D. willistoni* subspecies live in continental South America – *D. willistoni quechua* lives west of the Andes and *D. willistoni willistoni* east of the Andes. They are effectively separated by the Andes because the flies cannot live at high altitudes. It is not known whether their geographic separation is as old as the Andes, but it has existed long enough for postzygotic RIMs to have evolved. When the two subspecies are crossed in the laboratory, the hybrid males are completely sterile if the mother came from the *quechua* subspecies, but in the reciprocal cross all hybrids are fertile. If hybridization should occur in nature, selection would favour the evolution of prezygotic RIMs because of the complete sterility of half of the hybrid males.

Another pair of subspecies consists of *D. equinoxialis equinoxialis*, which inhabits continental South America, and *D. equinoxialis caribbensis*, which lives in Central America and the Caribbean. Crosses made in the laboratory between these two subspecies always produce sterile males, irrespective of the subspecies of the mother. Natural selection would, then, promote prezygotic RIMs between these two subspecies more strongly than between those of *D. willistoni*. But, in accord with the speciation model presented above, laboratory experiments show no evidence of the development of ethological isolation or of any other prezygotic RIM, presumably because the geographic isolation of the subspecies has forestalled hybridization between members.

One more sibling species of the group is *D. paulistorum*, a species that includes groups of populations well into the second stage of geographic speciation. Six such groups have been identified as semispecies, or incipient species, two or three of which are sympatric in many localities. Male hybrids between individuals of the different semispecies are sterile; laboratory crosses always yield fertile females but sterile males.

Whenever two or three incipient species of *D. paulistorum* have come into contact in nature, the second stage of speciation has led to the development of ethological isolation, which ranges from incipient to virtually complete. Laboratory experiments show that, when both incipient species are from the same locality, their ethological isolation is complete; only individuals of the same incipient species mate. When the individuals from different incipient species come from different localities, however, ethological isolation is usually present but far from complete. This is precisely as the speciation model predicts. Natural selection effectively promotes ethological isolation in territories where two incipient species live together, but the genes responsible for this isolation have not yet fully spread to populations in which one of the two incipient species is not present.

The eventual outcome of the process of geographic speciation is complete reproductive isolation, as can be observed among the species of the New World *Drosophila* group under discussion. *D. willistoni*, *D. equinoxialis*, *D. tropicalis*, and *D. paulistorum* coexist sympatrically over wide regions of Central and South America while preserving their separate gene pools. Hybrids are not known in nature and are almost impossible to obtain in the laboratory; moreover, all interspecific hybrid males at least are completely sterile. This total reproductive isolation has evolved, however, with very little morphological differentiation. Females from different sibling species cannot be distinguished by experts, while males can be identified only by small differences in the shape of their genitalia, unrecognizable except under a microscope.

Adaptive radiation

The geographic separation of populations derived from common ancestors may continue long enough that the populations become completely differentiated species before ever regaining sympatry and the opportunity to interbreed. As the allopatric populations continue evolving independently, RIMs develop and morphological differences may arise. The second stage of speciation – in which natural selection directly stimulates the evolution of RIMs – never comes about in such situations, because reproductive isolation takes place simply as a consequence of the continued separate evolution of the populations.

This form of allopatric speciation is particularly apparent when colonizers reach geographically remote areas, such as islands, where they find few or no competitors and have an opportunity to diverge as they become adapted to the new environment. Sometimes the new regions offer a multiplicity of environments to the colonizers, giving rise to several different lineages and species. This process of rapid divergence of multiple species from a single ancestral lineage is called adaptive radiation.

Many examples of speciation by adaptive radiation are found in archipelagoes removed from the mainland. The Galapagos islands are about 1,000 km (600 miles) off the west coast of South America. When Charles Darwin arrived there in 1835 during his voyage on the HMS *Beagle*, he discovered many species not found anywhere else in the world – for example, several species of finches, of which 14 are now known to exist (called Galapagos, or Darwin's, finches; Figure 5.3). These passerine birds have adapted to a diversity of habitats and diets, some feeding mostly on plants, others exclusively on insects. The various shapes of their bills are clearly adapted to probing, grasping, biting, or crushing – the diverse ways in which the different Galapagos species obtain their food. The explanation for such diversity is that the ancestor of Galapagos finches arrived in the islands before other kinds of birds and encountered an abundance of unoccupied ecological niches. Its descendants underwent adaptive radiation, evolving a variety of finch species with ways of life capable of exploiting opportunities that on various continents are already exploited by other species.

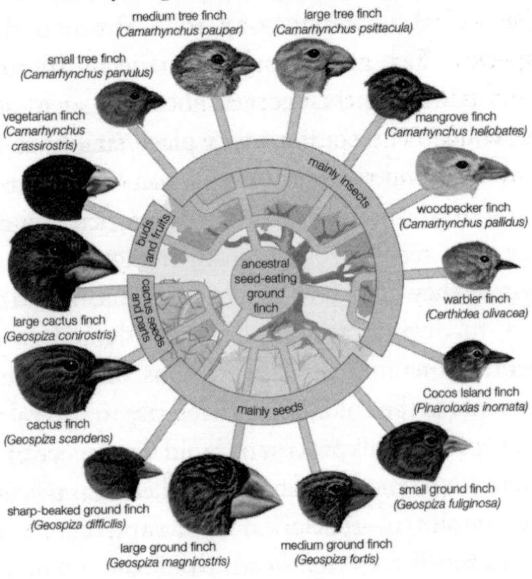

Figure 5.3 Adaptive radiation: 14 species of Galapagos finch that have evolved from a common ancestor.

The Hawaiian archipelago also provides striking examples of adaptive radiation. Its several volcanic islands, ranging from about one million to more than ten million years in age, are far from any continent or even other large islands. In their relatively small total land area, an astounding number of plant and animal species exist. Most of the species have evolved on the islands, among them about two dozen species (about one-third of them now extinct) of honeycreepers, birds of the family Drepanididae, all derived from a single immigrant form. In fact, all but one of Hawaii's 71 native bird species are endemic; that is, they have evolved there and are found nowhere else. More than 90 per cent of the native species of flowering plants, land molluscs, and insects are also endemic, as are two-thirds of the 168 species of ferns.

There are more than 500 native Hawaiian species of *Drosophila* flies – about one-third of the world's total number of known species. Far greater morphological and ecological diversity exists among the species in Hawaii than anywhere else in the world. The species of *Drosophila* in Hawaii have diverged by adaptive radiation from one or a few colonizers, which encountered an assortment of ecological niches that in other lands were occupied by different groups of flies or insects but that were available for exploitation in these remote islands.

Quantum speciation

In some modes of speciation the first stage is achieved in a short period of time. These modes are known by a variety of names, such as *quantum*, *rapid*, and *saltational* speciation, all suggesting the shortening of time involved. They are also known as *sympatric* speciation, alluding to the fact that quantum speciation often leads to speciation between populations that exist in the same territory or habitat. An important form of quantum speciation, polyploidy, is discussed separately below.

Quantum speciation without polyploidy has been seen in the annual plant genus *Clarkia*. Two closely related species, *Clarkia biloba* and *C. lingulata*, are both native to California. *C. lingulata* is known only from two sites in the central Sierra Nevada at the southern periphery of the distribution of *C. biloba*, from which it evolved starting with translocations and other chromosomal mutations (see above Chromosomal mutations). Such chromosomal rearrangements arise suddenly but reduce the fertility of heterozygous individuals. *Clarkia* species are capable of self-fertilization, which facilitates the propagation of the chromosomal mutants in different sets of individuals even within a single locality. This makes hybridization possible with non-mutant individuals and allows the second stage of speciation to go ahead.

Chromosomal mutations are often the starting point of quantum speciation in animals, particularly in groups such as mole rats and other rodents that live underground or have little mobility. Mole rats of the species group *Spalax ehrenbergi* in Israel and gophers of the species group *Thomomys talpoides* in the northern Rocky Mountains of the United States are well-studied examples.

The speciation process may also be initiated by changes in just one or a few gene loci when these alterations result in a change of ecological niche or, in the case of parasites, a change of host. Many parasites use their host as a place for courtship and mating, so organisms with two different host preferences may become reproductively isolated. If the hybrids show poor fitness because they are not effective parasites in either of the two hosts, natural selection will favour the development of additional RIMs. This type of speciation seems to be common among parasitic insects, a large group comprised of tens of thousands of species.

Polyploidy

As discussed above in Chromosomal mutations and in Chapter 4, the multiplication of entire sets of chromosomes is known as polyploidy. Whereas a diploid organism carries in the nucleus of each cell two sets of chromosomes, one inherited from each parent, a polyploid organism has three or more sets of chromosomes. Many cultivated plants are polyploid – bananas are triploid, potatoes are tetraploid, bread wheat is hexaploid, some strawberries are octaploid. These cultivated polyploids do not exist in nature, at least in any significant frequency. Some of them first appeared spontaneously; others, such as octaploid strawberries, were intentionally produced.

In animals polyploidy is relatively rare because it disrupts the balance between the sex chromosome and the other chromosomes, a balance that is required for the proper development of sex. Naturally polyploid species are found in hermaphroditic animals – individuals having both male and female organs – which include snails, earthworms, and planarians (a group of flatworms). They are also found in forms with parthenogenetic females (which produce viable progeny without fertilization), such as some beetles, sow bugs (a type of woodlouse), goldfish, and salamanders.

All major groups of plants have naturally polyploid species, but they are most common among angiosperms, or flowering plants, of which about 47 per cent are polyploids. Polyploidy is rare among gymnosperms, such as pines, firs, and cedars, although the redwood, *Sequoia sempervirens*, is a polyploid. Most polyploid plants are tetraploids. Polyploids with three, five, or some other odd-number multiple of the basic chromosome number are sterile, because the separation of homologous chromosomes cannot be achieved properly during formation of the sex cells. Some plants with an odd number of chromosome sets persist by means of asexual reproduction, particularly through human cultivation; the triploid banana is one example.

Polyploidy is a mode of quantum speciation that yields the beginnings of a new species in just one or two generations. There are two kinds of polyploids – autopolyploids, which derive from a single species, and allopolyploids, which stem from a combination of chromosome sets from different species. Allopolyploid plant species are much more numerous than autopolyploids.

An allopolyploid species can originate from two plant species that have the same diploid number of chromosomes. The chromosome complement of one species may be symbolized as *AA* and the other *BB*. A hybrid of two different species, represented as *AB*, will usually be sterile because of abnormal chromosome pairing and segregation during formation at meiosis of the gametes, which are haploid (i.e. having only half of the chromosomes, of which in a given gamete some come from the *A* set and some from the *B* set). But chromosome doubling may occur in a diploid cell as a consequence of abnormal mitosis, in which the chromosomes divide but the cell does not. If this happens in the hybrid above, *AB*, the result is a plant cell with four sets of chromosomes, *AABB*. Such a tetraploid cell may proliferate within the plant (which is otherwise constituted of diploid cells) and produce branches and flowers of tetraploid cells. Because the flowers' cells carry two chromosomes of each kind, they can produce functional diploid gametes via meiosis with the constitution *AB*. The union of two such gametes, such as happens during self-fertilization, produces a complete tetraploid individual (*AABB*). In this way, self-fertilization in plants makes possible the formation of a tetraploid individual as the result of a single abnormal cell division.

Autopolyploids originate in a similar fashion, except that the individual in which the abnormal mitosis occurs is not a hybrid. Self-fertilization thus enables a single individual to multiply and give rise to a population. This population is a new species, since polyploid individuals are reproductively isolated from their diploid ancestors. A cross between a tetraploid and a diploid yields triploid progeny, which are sterile.

Genetic Differentiation During Speciation

Genetic changes underlie all evolutionary processes. In order to understand speciation and its role in evolution, it is useful to know how much genetic change takes place during the course of species development. It is of considerable significance to ascertain whether new species arise by altering only a few genes or whether the process requires drastic changes – a genetic "revolution", as postulated by some evolutionists in the past. The issue is best considered separately with respect to each of the two stages of speciation and to the various modes of speciation.

The question of how much genetic differentiation occurs during speciation has become answerable only with the relatively recent development of appropriate methods for comparing genes of different species. Genetic change is measured with two parameters – genetic identity (I), which estimates the proportion of genes that are identical in two populations, and genetic distance (D), which estimates the proportion of gene changes that have occurred in the separate evolution of two populations. The value of I may range between 0 and 1, which corresponds to the extreme situations in which no or all genes are identical, respectively; the value of D may range from 0 to infinity. D can reach beyond 1 because each gene may change more than once in one or both populations as evolution goes on for many generations.

As a model of geographic speciation, the *Drosophila willistoni* group of flies offers the distinct advantage of exhibiting both stages of the speciation process. The *D. willistoni* group consists of several closely related species, some of which in turn consist of several incipient species, subspecies, or both. About 30 randomly selected genes have been studied in a large number of natural populations of these species. The results are summarized in Figure 5.4. The most significant numbers are those given in the levels of comparison labelled 2 and 3, which represent the first and second stages, respectively, of the process of geographic speciation. The 0.230 value for *D* (figure, level 2) means that about 23 gene changes have occurred for every 100 gene loci in the separate evolution of 2 subspecies – that is, the sum of the changes that have occurred in the 2 separately evolving lineages is 23 per cent of all the genes. These are populations well advanced in the first stage of speciation, as manifested by the sterility of the hybrid males.

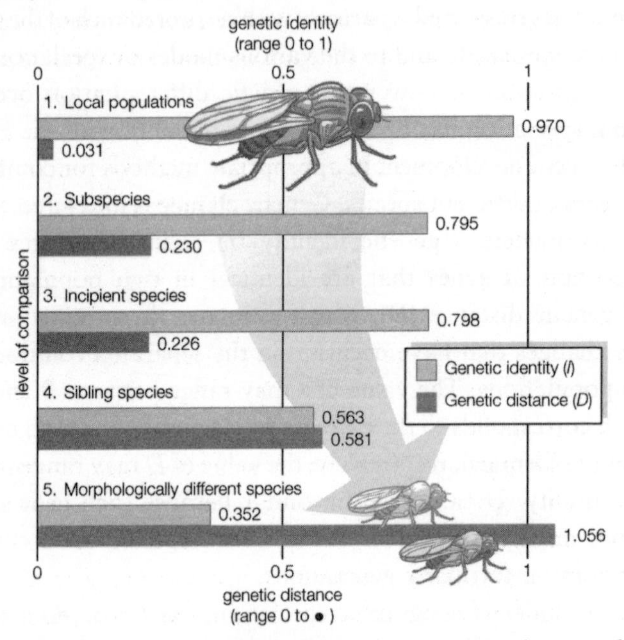

Figure 5.4 Genetic differentiation between populations of *Drosophila willistoni.*

The genetic distance between incipient species (figure, level 3) is the same, within experimental error, as that between the subspecies, or 22.6 per cent. This implies that the development of ethological isolation, as it is found in these populations, does not require many genetic changes beyond those that occurred during the first stage of speciation. Indeed, no additional gene changes were detected in these experiments. The absence of major genetic changes during the second stage of speciation can be understood by considering the role of natural selection, which directly promotes the evolution of prezygotic RIMs during the second stage, so that only genes modifying mate choice need to change. In contrast, the development of postzygotic RIMs during the first stage occurs only after there is substantial genetic differentiation between populations, because it comes about only as an incidental outcome of overall genetic divergence.

Sibling species, such as *D. willistoni* and *D. equinoxialis*, exhibit 58 gene changes for every 100 gene loci after their divergence from a common ancestor (figure, level 4). It is noteworthy that this much genetic evolution has occurred without altering the external morphology of these organisms. In the evolution of morphologically different species (figure, level 5), the number of gene changes is greater yet, as would be expected.

Genetic changes concomitant with one or the other of the two stages in the speciation process have been studied in a number of organisms, from insects and other invertebrates to all sorts of vertebrates, including mammals. The amount of genetic change during geographic speciation varies between organisms, but the two main observations made in the *D. willistoni* group seem to apply quite generally. These are that the evolution of postzygotic mechanisms during the first stage is accompanied by substantial genetic change (a majority of values for genetic distance, *D*, range between 0.15 and 0.30) and that relatively few additional genetic changes are required during the second stage.

The conclusions drawn from the investigation of geographic speciation make it possible to predict the relative amounts of genetic change expected in the quantum modes of speciation. Polyploid species are a special case – they arise suddenly in one or a few generations, and at first they are not expected to be genetically different from their ancestors. More generally, quantum speciation involves a shortening of the first stage of speciation, so that postzygotic RIMs arise directly as a consequence of specific genetic changes (such as chromosome mutations). Populations in the first stage of quantum speciation, therefore, need not be substantially different in individual gene loci. This has been confirmed by genetic investigations of species recently arisen by quantum speciation. For example, the average genetic distance between four incipient species of the mole rat *Spalax ehrenbergi* is 0.022, and between those of the gopher *Thomomys talpoides* it is 0.078. The second stage of speciation is modulated in essentially the same way as in the geographic mode. Not many gene changes are needed in either case to complete speciation.

Patterns and Rates of Species Evolution

Evolution Within a Lineage and by Lineage Splitting

Evolution can take place by anagenesis, in which changes occur within a lineage, or by cladogenesis, in which a lineage splits into two or more separate lines. Anagenetic evolution has doubled the size of the human cranium over the course of two million years; in the lineage of the horse it has reduced the number of toes from four to one. Cladogenetic evolution has produced the extraordinary diversity of the living world, with its more than two million species of animals, plants, fungi, and microorganisms.

The most essential cladogenetic function is speciation, the process by which one species splits into two or more species.

Because species are reproductively isolated from one another, they are independent evolutionary units; that is, evolutionary changes occurring in one species are not shared with other species. Over time, species diverge more and more from one another as a consequence of anagenetic evolution. Descendant lineages of two related species that existed millions of years ago may now be classified into quite different biological categories, such as different genera or even different families.

The evolution of all living organisms, or of a subset of them, can be seen as a tree, with branches that divide into two or more as time progresses. Such trees are called phylogenies. Their branches represent evolving lineages, some of which eventually die out while others persist in themselves or in their derived lineages down to the present time. Evolutionists are interested in the history of life and hence in the topology, or configuration, of phylogenies. They are concerned as well with the nature of the anagenetic changes within lineages and with the timing of the events.

Phylogenetic relationships are ascertained by means of several complementary sources of evidence. First, there are the discovered remnants of organisms that lived in the past, the fossil record, which provides definitive evidence of relationships between some groups of organisms. The fossil record, however, is far from complete and is often seriously deficient. Second, information about phylogeny comes from comparative studies of living forms. Comparative anatomy contributed the most information in the past, although additional knowledge came from comparative embryology, cytology, ethology, biogeography, and other biological disciplines. In recent years the comparative study of the so-called informational macromolecules – proteins and nucleic acids, whose specific sequences of constituents carry genetic information – has become a powerful tool for the study of phylogeny (see below

DNA and Protein as Informational Macromolecules, and see also Molecular Biology earlier in this chapter).

The modern theory of evolution provides a causal explanation of the similarities between living things. Organisms evolve by a process of descent with modification. Changes, and therefore differences, gradually accumulate over the generations. The more recent the last common ancestor of a group of organisms, the less their differentiation; similarities of form and function reflect phylogenetic propinquity. Accordingly, phylogenetic affinities can be inferred on the basis of relative similarity.

Convergent and Parallel Evolution

A distinction has to be made between resemblances due to propinquity of descent and those due only to similarity of function. Structural similarity in different organisms that is due to inheritance from a common ancestor is called homology. The forelimbs of humans, whales, dogs, and bats are homologous. The skeletons of these limbs are all constructed of bones arranged according to the same pattern because they derive from a common ancestor with similarly arranged forelimbs. Correspondence of features due to similarity of function but not related to common descent is termed analogy. The wings of birds and of flies are analogous. Their wings are not modified versions of a structure present in a common ancestor but rather have developed independently as adaptations to a common function, flying. The similarities between the wings of bats and birds are partially homologous and partially analogous. Their skeletal structure is homologous, due to common descent from the forelimb of a reptilian ancestor; but the modifications for flying are different and independently evolved, and in this respect they are analogous.

Features that become more rather than less similar through independent evolution are said to be convergent. Convergence is often associated with similarity of function, as in the evolution of wings in birds, bats, and flies. The shark (a fish) and the dolphin (a mammal) are much alike in external morphology; their similarities are due to convergence, since they have evolved independently as adaptations to aquatic life.

Taxonomists also speak of parallel evolution. Parallelism and convergence are not always clearly distinguishable. Strictly speaking, convergent evolution occurs when descendants resemble each other more than their ancestors did with respect to some feature. Parallel evolution implies that two or more lineages have changed in similar ways, so that the evolved descendants are as similar to each other as their ancestors were. The evolution of marsupials in Australia, for example, paralleled the evolution of placental mammals in other parts of the world. There are Australian marsupials resembling true wolves, cats, mice, squirrels, moles, groundhogs, and anteaters. These placental mammals and the corresponding Australian marsupials evolved independently but in parallel lines by reason of their adaptation to similar ways of life. Some resemblances between a true anteater (genus *Myrmecophaga*) and a marsupial anteater, or numbat (*Myrmecobius*), are due to homology – both are mammals. Others are due to analogy – both feed on ants.

Parallel and convergent evolution are also common in plants. New World cacti and African euphorbias, or spurges, are alike in overall appearance although they belong to separate families. Both are succulent, spiny, water-storing plants adapted to the arid conditions of the desert. Their corresponding morphologies have evolved independently in response to similar environmental challenges.

Homology can be recognized not only between different organisms but also between repetitive structures of the same organism. This has been called serial homology. There is serial homology, for example, between the arms and legs of humans, between the seven cervical vertebrae of mammals, and between the branches or leaves of a tree. The jointed appendages of arthropods are elaborate examples of serial homology. Crayfish have 19 pairs of appendages, all built according to the same basic pattern but serving diverse functions – sensing, chewing, food handling, walking, mating, egg carrying, and swimming. Although serial homologies are not useful in reconstructing the phylogenetic relationships of organisms, they are an important dimension of the evolutionary process.

Relationships in some sense akin to those between serial homologues exist at the molecular level between genes and proteins derived from ancestral gene duplications. Consider the iron-containing protein haemoglobin, which is made up of four polypeptide chains, each chain consisting of more than 140 amino acids. Normal adult human haemoglobin consists of two alpha and two beta chains. About 500 million years ago a chromosome segment carrying the gene coding for haemoglobin became duplicated, so that the genes in the different segments thereafter evolved in somewhat different ways, one eventually giving rise to the modern gene coding for the alpha haemoglobin chain, the other for the beta chain. The beta chain gene became duplicated again about 200 million years ago, giving rise to the gamma haemoglobin chain, a normal component of fetal haemoglobin (haemoglobin F). The genes for the alpha, beta, gamma, and other haemoglobin chains are homologous; similarities in their nucleotide sequences occur because they are modified descendants of a single ancestral sequence.

There are two ways of comparing homology between haemoglobins. One is to compare the same haemoglobin chain – for instance, the alpha chain – in different species of animals. The degree of divergence between the alpha chains reflects the degree of the evolutionary relationship between the organisms, because the haemoglobin chains have evolved independently of one another since the time of divergence of the lineages leading to the present-day organisms. A second way is to make comparisons between, say, the alpha and beta chains of a single species. The degree of divergence between the different globin chains reflects the degree of relationship between the genes coding for them. The different globins have evolved independently of each other since the time of duplication of their ancestral genes. Comparisons between homologous genes or proteins within a given organism provide information about the phylogenetic history of the genes and hence about the historical sequence of the gene duplication events.

Whether similar features in different organisms are homologous or analogous – or simply accidental – cannot always be decided unambiguously, but the distinction must be made in order to determine phylogenetic relationships. Moreover, the degrees of homology must be quantified in some way so as to determine the propinquity of common descent between species. Difficulties arise here as well. In the case of forelimbs, it is not clear whether the homologies are greater between human and bird than between human and reptile, or between human and reptile than between human and bat. The fossil record, though deficient, sometimes provides the appropriate information. Fossil evidence must be examined together with the evidence from comparative studies of living forms and with the quantitative estimates provided by comparative studies of proteins and nucleic acids.

Gradual and Punctuational Evolution

The fossil record indicates that morphological evolution is by and large a gradual process. Major evolutionary changes are usually due to a building-up over the ages of relatively small changes. But the fossil record is discontinuous. Fossil strata are separated by sharp boundaries; accumulation of fossils within a geologic deposit (stratum) is fairly constant over time, but the transition from one stratum to another may involve gaps of tens of thousands of years. Whereas the fossils within a stratum exhibit little morphological variation, new species – characterized by small but discontinuous morphological changes – typically appear at the boundaries between strata. That is not to say that the transition from one stratum to another always involves sudden changes in morphology; on the contrary, fossil forms often persist virtually unchanged through several geologic strata, each representing millions of years.

The apparent morphological discontinuities of the fossil record are often attributed by palaeontologists to the discontinuity of the sediments – that is, to the substantial time gaps encompassed in the boundaries between strata. The assumption is that, if the fossil deposits were more continuous, they would show a more gradual transition of form. Even so, morphological evolution would not always keep progressing gradually, because some forms, at least, remain unchanged for extremely long times. Examples are the lineages known as "living fossils" – for instance, the lamp shell *Lingula*, a genus of brachiopod (a phylum of shelled invertebrates) that appears to have remained essentially unchanged since the Ordovician Period, some 450 million years ago; or the tuatara, *Sphenodon punctatus*, a reptile that has shown little morphological evolution for nearly 200 million years, since the early Mesozoic Era.

Some palaeontologists have proposed that the discontinuities of the fossil record are not artefacts created by gaps in the record but rather reflect the true nature of morphological evolution, which happens in sudden bursts associated with the formation of new species. The lack of morphological evolution, or stasis, of lineages such as *Lingula* and *Sphenodon* is in turn due to lack of speciation within those lineages. The proposition that morphological evolution is jerky, with most morphological change occurring during the brief speciation events and virtually no change during the subsequent existence of the species, is known as the punctuated equilibrium model.

Whether morphological evolution in the fossil record is predominantly punctuational or gradual is a much-debated question. The imperfection of the record makes it unlikely that the issue will be settled in the foreseeable future. Intensive study of a favourable and abundant set of fossils may be expected to substantiate punctuated or gradual evolution in particular cases. But the argument is not about whether only one or the other pattern ever occurs; it is about their relative frequency. Some palaeontologists argue that morphological evolution is in most cases gradual and only rarely jerky, whereas others think the opposite is true.

The proponents of the punctuated equilibrium model propose not only that morphological evolution is jerky but also that it is associated with speciation events. They argue that phyletic evolution – that is, evolution along lineages of descent – proceeds at two levels. First, there is continuous change through time within a population. This consists largely of gene substitutions prompted by natural selection, mutation, genetic drift, and other genetic processes that operate at the level of the individual organism. The punctualists maintain that this continuous evolution within established lineages rarely, if ever, yields substantial morphological changes in species. Second, they say, there is the process of origination and extinction of species, in which most morphological change occurs. According to the punctualist model, evolutionary trends result from the patterns of origination and extinction of species rather than from evolution within established lineages.

Diversity and Extinction

The current diversity of life is the balance between the species that have arisen through time and those that have become extinct. Palaeontologists observe that organisms have continuously changed since the Cambrian period, more than 500 million years ago, from which abundant animal fossil remains are known. The division of geologic history into a succession of eras and periods is hallmarked by major changes in plant and animal life – the appearance of new sorts of organisms and the extinction of others. Palaeontologists distinguish between background extinction, the steady rate at which species disappear through geologic time, and mass extinctions, the episodic events in which large numbers of species become extinct over time spans short enough to appear almost instantaneous on the geologic scale.

Best known among mass extinctions is the one that occurred at the end of the Cretaceous period, when the dinosaurs and many other marine and land animals disappeared. Most scientists believe that the Cretaceous mass extinction was provoked by the impact of an asteroid or comet on the tip of the Yucatán peninsula in south-eastern Mexico 65 million years ago. The object's impact caused an enormous dust cloud, which greatly reduced the amount of solar radiation reaching the Earth, with a consequent drastic drop in temperature and other adverse conditions. Among animals, about 76 per cent of species, 47 per cent of genera, and 16 per cent of families became extinct. Although the dinosaurs vanished, turtles, snakes, lizards, crocodiles, and other reptiles, as well as some mammals and birds, survived. Mammals that lived before this event were small and mostly nocturnal, but during the ensuing Tertiary period they experienced an explosive diversification in size and morphology, occupying ecological niches vacated by the dinosaurs. Most of the orders and families of mammals now in existence originated in the first 10 million–20 million years after the dinosaurs' extinction. Birds also greatly diversified at that time.

Several other mass extinctions have occurred since the Cambrian. The most catastrophic happened at the end of the Permian period, about 248 million years ago, when 95 per cent of species, 82 per cent of genera, and 51 per cent of families of animals became extinct. Other large mass extinctions occurred at or near the end of the Ordovician (about 440 million years ago, 85 per cent of species extinct), Devonian (about 360 million years ago, 83 per cent of species extinct), and Triassic (about 210 million years ago, 80 per cent of species extinct). Changes of climate and chemical composition of the atmosphere appear to have caused these mass extinctions; there is no convincing evidence that they resulted from cosmic impacts. Like other mass extinctions, they were followed by the origin or rapid diversification of various kinds of organisms. The first mammals and dinosaurs appeared after the late Permian extinction, and the first vascular plants after the Late Ordovician extinction.

Background extinctions result from ordinary biological processes, such as competition between species, predation, and parasitism. When two species compete for very similar resources – say, the same kinds of seeds or fruits – one may become extinct, although often they will displace one another by dividing the territory or by specializing in slightly different foods, such as seeds of a different size or kind. Ordinary physical and climatic changes also account for background extinctions – for example, when a lake dries out or a mountain range rises or erodes.

New species come about by the processes discussed in previous sections. These processes are largely gradual, yet the history of life shows major transitions in which one kind of organism becomes a very different kind. The earliest organisms were prokaryotes, or bacteria-like cells, whose hereditary material is not segregated into a nucleus. Eukaryotes have their DNA organized into chromosomes that are membrane-bound in the nucleus, have other organelles inside their cells, and reproduce sexually. Eventually, eukaryotic multicellular organisms appeared, in which there is a division of function among cells – some specializing in reproduction, others becoming leaves, trunks, and roots in plants or different organs and tissues such as muscle, nerve, and bone in animals. Social organization of individuals in a population is another way of achieving functional division, which may be quite fixed, as in ants and bees, or more flexible, as in herds of cattle or groups of primates.

Because of the gradualness of evolution, immediate descendants differ little, and then mostly quantitatively, from their ancestors. But gradual evolution may amount to large differences over time. The forelimbs of mammals are normally adapted for walking, but they are adapted for shovelling earth in moles and other mammals that live mostly underground; for climbing and grasping in arboreal monkeys and apes; for swimming in dolphins and whales; and for flying in bats. The forelimbs of reptiles became wings in their bird descendants. Feathers appear to have served first for regulating temperature but eventually were co-opted for flying and became incorporated into wings.

Eyes also evolved gradually and achieved very different configurations, all serving the function of seeing. Eyes have evolved independently at least 40 times. Because sunlight is a pervasive feature of Earth's environment, it is not surprising that organs have evolved that take advantage of it. The simplest "organ" of vision occurs in some single-celled organisms that have enzymes or spots sensitive to light, which help them move towards the surface of their pond, where they feed on the algae growing there by photosynthesis. Some multicellular animals exhibit light-sensitive spots on their epidermis. Further steps – deposition of pigment around the spot, configuration of cells into a cup-like shape, thickening of the epidermis leading to the development of a lens, development of muscles to move the eyes and nerves to transmit optical signals to the brain – all led to the highly developed eyes of vertebrates and cephalopods (octopuses and squids) and to the compound eyes of insects.

While the evolution of forelimbs – for walking – into the wings of birds or the arms and hands of primates may seem more like changes of function, the evolution of eyes exemplifies gradual advancement of the same function – seeing. In all cases, however, the process is impelled by natural selection's favouring individuals exhibiting functional advantages over others of the same species. Examples of functional shifts are many and diverse. Some transitions may at first seem unlikely because of the difficulty in identifying which possible functions may have been served during the intermediate stages. These cases are eventually resolved with further research and the discovery of intermediate fossil forms. An example of a seemingly unlikely transition is the transformation of bones found in the reptilian jaw into the hammer and anvil of the mammalian ear.

Evolution and Development

Starfish are radially symmetrical, but most animals are bilaterally symmetrical – the parts of the left and right halves of their bodies tend to correspond in size, shape, and position. Some bilateral animals, such as millipedes and shrimps, are segmented (metameric); others, such as frogs and humans, have a front-to-back (head-to-foot) body plan, with head, thorax, abdomen, and limbs, but they lack the repetitive, nearly identical segments of metameric animals. There are other basic body plans, such as those of sponges, clams, and jellyfish, but their total number is not large – less than 40.

The fertilized egg, or zygote, is a single cell, more or less spherical, that does not exhibit polarity such as anterior and posterior ends or dorsal and ventral sides. Embryonic development is the process of growth and differentiation by which the single-celled egg becomes a multicellular organism.

The determination of body plan from this single cell and the construction of specialized organs, such as the eye, are under the control of regulatory genes. Most notable among these are the *Hox* genes, which produce proteins (transcription factors) that bind with other genes and thus determine their expression – that is, when they will act. The *Hox* genes embody spatial and temporal information. By means of their encoded proteins, they activate or repress the expression of other genes according to the position of each cell in the developing body, determining where limbs and other body parts will grow in the embryo. Since their discovery in the early 1980s, the *Hox* genes have been found to play crucial roles from the first steps of development, such as establishing anterior and posterior ends in the zygote, to much later steps, such as the differentiation of nerve cells.

The critical region of the *Hox* proteins is encoded by a sequence of about 180 consecutive nucleotides. The corresponding protein region, about 60 amino acids long, binds to a short stretch of DNA in the regulatory region of the target genes. Genes containing similar sequences are found not only in animals but also in other eukaryotes such as fungi and plants.

All animals have *Hox* genes; there may be as few as 1, as in sponges, or as many as 38, as in humans and other mammals. *Hox* genes are clustered in the genome. Invertebrates have only 1 cluster with a variable number of genes, typically fewer than 13. The common ancestor of the chordates (which includes the vertebrates) probably had only 1 cluster of *Hox* genes, which may have numbered 13. Chordates may have 1 or more clusters, but not all 13 genes remain in every cluster. The marine animal *Amphioxus*, a primitive chordate, has a single array of 10 *Hox* genes. Humans, mice, and other mammals have 38 *Hox* genes arranged in 4 clusters, 3 with 9 genes each and 1 with 11 genes. The set of genes varies from cluster to cluster, so that out of the 13 in the original cluster, genes designated 1, 2, 3, and 7 may be missing in one set, whereas 10, 11, 12, and 13 may be missing in a different set.

The 4 clusters of *Hox* genes found in mammals originated by duplication of the whole original cluster and retain considerable similarity between clusters. The 13 genes in the original cluster also themselves originated by repeated duplication, starting from a single *Hox* gene as found in the sponges. These first duplications happened very early in animal evolution, in the Precambrian. The genes within a cluster retain detectable similarity, but they differ more from one another than they differ from the corresponding, or homologous, gene in any of the other sets. There is a puzzling correspondence between the position of the *Hox* genes in a cluster along the chromosome and the patterning of the body – genes located upstream (anteriorly in the direction in which genes are transcribed) in the cluster are expressed earlier and more anteriorly in the body, while those located downstream (posteriorly in the direction of transcription) are expressed later in development and predominantly affect the posterior body parts.

Researchers demonstrated the evolutionary conservation of the *Hox* genes by means of clever manipulations of genes in laboratory experiments. For example, the *ey* gene that determines the formation of the compound eye in *Drosophila* vinegar flies was activated in the developing embryo in various parts of the body, yielding experimental flies with anatomically normal eyes on the legs, wings, and other structures. The evolutionary conservation of the *Hox* genes may be the explanation for the puzzling observation that most of the diversity of body plans within major groups of animals arose early in the evolution of the group. The multicellular animals (metazoans) first found as fossils in the Cambrian already demonstrate all the major body plans found during the ensuing 540 million years, as well as 4 to 7 additional body plans that became extinct and seem bizarre to observers today. Similarly, most of the classes found within a phylum appear early in the evolution of the phylum. For example, all living classes of arthropods are already found in the Cambrian, with body plans essentially unchanged thereafter; in addition, the Cambrian contains a few strange kinds of arthropods that later became extinct (Figure 5.5).

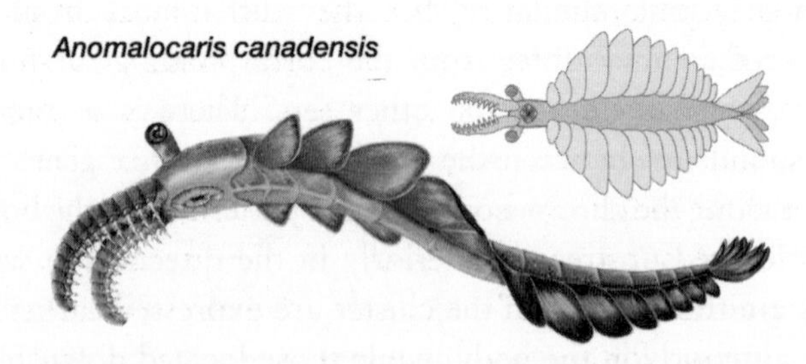

Figure 5.5 Sketch of *Anomalocaris canadensis*. Members of the genus *Anomalocaris* were the largest marine predators of the Cambrian period.

Reconstruction of Evolutionary History

DNA and Protein as Informational Macromolecules

The advances of molecular biology have made possible the comparative study of proteins and the nucleic acids, DNA and RNA. DNA is the repository of hereditary (evolutionary and developmental) information. The relationship of proteins to DNA is so immediate that they closely reflect the hereditary information. This reflection is not perfect, because the genetic code is redundant, and, consequently, some differences in the DNA do not yield differences in the proteins. Moreover, this reflection is not complete, because a large fraction of DNA (about 90 per cent in many organisms) does not code for proteins. Nevertheless, proteins are so closely related to the information contained in DNA that they, as well as nucleic acids, are called informational macromolecules.

Nucleic acids and proteins are linear molecules made up of sequences of units – nucleotides in the case of nucleic acids, amino acids in the case of proteins – which retain considerable amounts of evolutionary information. Comparing two macromolecules establishes the number of their units that are different. Because evolution usually occurs by changing one unit at a time, the number of differences is an indication of the recency of common ancestry. Changes in evolutionary rates may create difficulties in interpretation, but macromolecular studies have three notable advantages over comparative anatomy and the other classical disciplines. One is that the information is more readily quantifiable. The number of units that are different is readily established when the sequence of units is known for a given macromolecule in different organisms. The second advantage is that comparisons can be made even between very different sorts of organisms. There is very little that comparative anatomy can say when organisms as diverse as yeasts, pine trees, and human beings are compared, but there are homologous macromolecules that can be compared in all three. The third advantage is multiplicity. Each organism possesses thousands of genes and proteins, which all reflect the same evolutionary history. If the investigation of one particular gene or protein does not resolve the evolutionary relationship of a set of species, additional genes and proteins can be investigated until the matter has been settled.

Informational macromolecules provide information not only about the branching of lineages from common ancestors (cladogenesis) but also about the amount of genetic change that has occurred in any given lineage (anagenesis). It might seem at first that quantifying anagenesis for proteins and nucleic acids would be impossible, because it would require comparison of molecules from organisms that lived in the past with those from living organisms. Organisms of the past are sometimes preserved as fossils, but their DNA and proteins have largely disintegrated. Nevertheless, comparisons between living species provide information about anagenesis.

As a concrete example, consider the protein cytochrome c, involved in cell respiration. The sequence of amino acids in this protein is known for many organisms, from bacteria and yeasts to insects and humans; in animals cytochrome c consists of 104 amino acids. When the amino acid sequences of humans and rhesus monkeys are compared, they are found to be different at position 66 (isoleucine in humans, threonine in rhesus monkeys) but identical at the other 103 positions. When humans are compared with horses, 12 amino acid differences are found, but, when horses are compared with rhesus monkeys, there are only 11 amino acid differences. Even without knowing anything else about the evolutionary history of mammals, one would conclude that the lineages of humans and rhesus monkeys diverged from each other much more recently than they diverged from the horse lineage. Moreover, it can be concluded that the amino acid difference between humans and rhesus monkeys must have occurred in the human lineage after its separation from the rhesus monkey lineage (see Figure 4.4).

Evolutionary Trees

Evolutionary trees are models that seek to reconstruct the evolutionary history of taxa – i.e. species or other groups of organisms, such as genera, families, or orders. The trees embrace two kinds of information related to evolutionary change, cladogenesis and anagenesis. Figure 4.4 illustrates both kinds. The branching relationships of the trees reflect the relative relationships of ancestry, or cladogenesis.

Evolutionary trees may also indicate the changes that have occurred along each lineage, or anagenesis. Thus, in the evolution of cytochrome c since the last common ancestor of humans and rhesus monkeys (the right side of the figure), one amino acid changed in the lineage going to humans but none in the lineage going to rhesus monkeys. Similarly, the left side of the figure shows that three amino acid changes occurred in the lineage from B to C but only one in the lineage from B to D.

There are several methods for constructing evolutionary trees. Some were developed for interpreting morphological data, others for interpreting molecular data; some can be used with either kind of data. Figure 5.6 demonstrates an evolutionary tree based on the minimum number of nucleotide differences in the genes of 20 species that are necessary to account for the differences in their cytochome c.

The relationships between species as shown in the figure correspond fairly well to the relationships determined from other sources, such as the fossil record. According to the

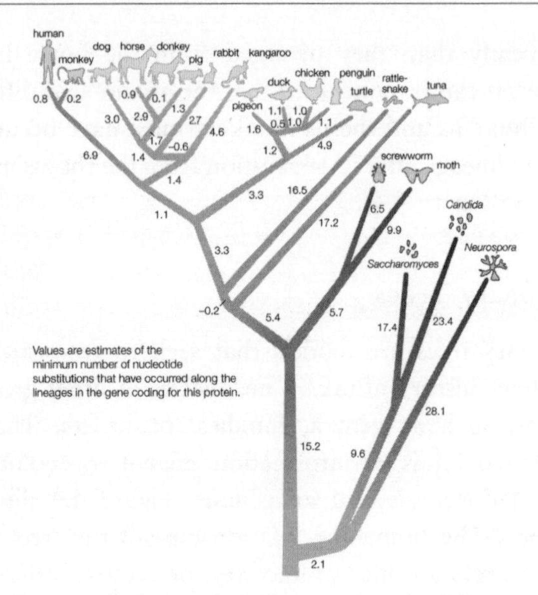

Figure 5.6 Phylogeny based on nucleotide differences for cytochrome c.

figure, chickens are less closely related to ducks and pigeons than to penguins, and humans and monkeys diverged from the other mammals before the marsupial kangaroo separated from the non-primate placentals. Although these examples are known to be erroneous relationships, the power of the method is apparent in that a single protein yields a fairly accurate reconstruction of the evolutionary history of 20 organisms that started to diverge more than one billion years ago.

Molecular Evolution

Molecular Evolution of Genes

The methods for obtaining the nucleotide sequences of DNA have enormously improved since the 1980s and have become largely automated. Many genes have been sequenced in numerous organisms, and the complete genome has been sequenced in various species ranging from humans to viruses. The use of DNA sequences has been particularly rewarding in the study of gene duplications. The genes that code for the haemoglobins in humans and other mammals provide a good example.

Knowledge of the amino acid sequences of the haemoglobin chains and of myoglobin, a closely related protein, has made it possible to reconstruct the evolutionary history of the duplications that gave rise to the corresponding genes. But direct examination of the nucleotide sequences in the genes coding for these proteins has shown that the situation is more complex, and also more interesting, than it appears from the protein sequences. DNA sequence studies on human haemoglobin genes have shown that their number is greater than previously thought. The similarity in the nucleotide sequence indicates that they have arisen through various duplications and subsequent evolution from a gene ancestral to all.

Multiplicity and Rate Heterogeneity

Cytochrome c consists of only 104 amino acids, encoded by 312 nucleotides. Nevertheless, this short protein stores enormous evolutionary information, which made possible the fairly good approximation to the evolutionary history of 20 very diverse species over a period longer than 1 billion years (see Figure 5.6).

But cytochrome c is a slowly evolving protein. Widely different species have in common a large proportion of the amino acids in their cytochrome c, which makes possible the study of genetic differences between organisms only remotely related. For the same reason, however, comparing cytochrome c molecules cannot determine evolutionary relationships between closely related species. For example, the amino acid sequence of cytochrome c in humans and chimpanzees is identical, although they diverged about 6 million years ago; between humans and rhesus monkeys, which diverged from their common ancestor 35 million to 40 million years ago, it differs by only a single amino acid replacement.

Proteins that evolve more rapidly than cytochrome c can be studied in order to establish phylogenetic relationships between closely related species. Some proteins evolve very fast; the fibrinopeptides – small proteins involved in the blood-clotting process – are suitable for reconstructing the phylogeny of recently evolved species, such as closely related mammals. Other proteins evolve at intermediate rates; the haemoglobins, for example, can be used for reconstructing evolutionary history over a fairly broad range of time.

One great advantage of molecular evolution is its multiplicity, as noted above in the section on DNA and Protein as Informational Macromolecules. Within each organism are thousands of genes and proteins; these evolve at different rates, but every one of them reflects the same evolutionary events. Scientists can obtain greater and greater accuracy in reconstructing the evolutionary phylogeny of any group of organisms by increasing the number of genes investigated. The range of differences in the rates of evolution between genes opens up the opportunity of investigating different sets of genes for achieving different degrees of resolution in the tree, relying on slowly evolving ones for remote evolutionary events. Even genes that encode slowly evolving proteins can be useful for reconstructing the evolutionary relationships between closely related species, by examination of the redundant codon substitutions (nucleotide substitutions that do not change the encoded amino acids), the introns (non-coding

DNA segments interspersed among the segments that code for amino acids), or other non-coding segments of the genes (such as the sequences that precede and follow the encoding portions of genes); these generally evolve much faster than the nucleotides that specify the amino acids.

Molecular Clock of Evolution

One conspicuous attribute of molecular evolution is that differences between homologous molecules can readily be quantified and expressed, as, for example, proportions of nucleotides or amino acids that have changed. Rates of evolutionary change can therefore be more precisely established with respect to DNA or proteins than with respect to phenotypic traits of form and function. Studies of molecular evolution rates have led to the proposition that macromolecules may serve as evolutionary clocks.

It was first observed in the 1960s that the numbers of amino acid differences between homologous proteins of any two given species seemed to be nearly proportional to the time of their divergence from a common ancestor. If the rate of evolution of a protein or gene were approximately the same in the evolutionary lineages leading to different species, proteins and DNA sequences would provide a molecular clock of evolution. The sequences could then be used to reconstruct not only the sequence of branching events of a phylogeny but also the time when the various events occurred.

Consider, for example, Figure 5.6, depicting the 20-organism phylogeny. If the substitution of nucleotides in the gene coding for cytochrome c occurred at a constant rate through time, one could determine the time elapsed along any branch of the phylogeny simply by examining the number of nucleotide substitutions along that branch. One would need only to calibrate the clock by reference to an outside source, such as the fossil record, that would provide the actual geologic time elapsed in at least one specific lineage.

The molecular evolutionary clock, of course, is not expected to be a metronomic clock, like a watch or other timepiece that measures time exactly, but a stochastic clock like radioactive decay. In a stochastic clock the probability of a certain amount of change is constant (for example, a given quantity of atoms of radium-226 is expected, through decay, to be reduced by half in 1,620 years), although some variation occurs in the actual amount of change. Over fairly long periods of time a stochastic clock is quite accurate. The enormous potential of the molecular evolutionary clock lies in the fact that each gene or protein is a separate clock. Each clock "ticks" at a different rate – the rate of evolution characteristic of a particular gene or protein – but each of the thousands and thousands of genes or proteins provides an independent measure of the same evolutionary events.

Evolutionists have found that the amount of variation observed in the evolution of DNA and proteins is greater than is expected from a stochastic clock – in other words, the clock is erratic. The discrepancies in evolutionary rates along different lineages are not excessively large, however. So it is possible, in principle, to time phylogenetic events with as much accuracy as may be desired, but more genes or proteins (about two to four times as many) must be examined than would be required if the clock was stochastically constant. The average rates obtained for several proteins taken together become a fairly precise clock, particularly when many species are studied and the evolutionary events involve long time periods (on the order of 50 million years or longer). The more recent the divergence of any two species, the more likely it is that the changes observed will depart from the average evolutionary rate. As the length of time increases, periods of rapid and slow evolution in any lineage are likely to cancel one another out.

Evolutionists have discovered, however, that molecular time estimates tend to be systematically older than estimates based on other methods and, indeed, to be older than the actual dates. This is a consequence of the statistical properties of molecular estimates, which are asymmetrically distributed. Because of chance, the number of molecular differences between two species may be larger or smaller than expected. But overestimation errors are unbounded, whereas underestimation errors are bounded, since they cannot be smaller than 0. Consequently, a graph of a typical distribution of estimates of the age when two species diverged, gathered from a number of different genes, is skewed from the normal bell shape, with a large number of estimates of younger age clustered together at one end and a long "tail" of older-age estimates trailing away toward the other end. The average of the estimated times thus will consistently overestimate the true date. The overestimation bias becomes greater when the rate of molecular evolution is slower, the sequences used are shorter, and the time becomes increasingly remote.

Neutrality Theory of Molecular Evolution

In the late 1960s it was proposed that at the molecular level most evolutionary changes are selectively "neutral", meaning that they are due to genetic drift rather than to natural selection. Nucleotide and amino acid substitutions appear in a population by mutation. If alternative alleles (alternative DNA sequences) have identical fitness – if they are identically able to perform their function – changes in allelic frequency from generation to generation will occur only by genetic drift.

Rates of allelic substitution will be stochastically constant – that is, they will occur with a constant probability for a given gene or protein. This constant rate is the mutation rate for neutral alleles.

According to the neutrality theory, a large proportion of all possible mutants at any gene locus are harmful to their carriers. These mutants are eliminated by natural selection, just as standard evolutionary theory postulates. The neutrality theory also agrees that morphological, behavioral, and ecological traits evolve under the control of natural selection. What is distinctive in the theory is the claim that at each gene locus there are several favourable mutants, equivalent to one another with respect to adaptation, so that they are not subject to natural selection among themselves. Which of these mutants increases or decreases in frequency in one or another species is purely a matter of chance, the result of random genetic drift over time.

Neutral alleles are those that differ so little in fitness that their frequencies change by random drift rather than by natural selection. It can be theorized that the rate of substitution of neutral alleles is precisely the rate at which the neutral alleles arise by mutation, independently of the number of individuals in the population or of any other factors.

If the neutrality theory of molecular evolution is strictly correct, it will provide a foundation for the hypothesis of the molecular evolutionary clock, since the rate of neutral mutation would be expected to remain constant through evolutionary time and in different lineages. The number of amino acid or nucleotide differences between species would, therefore, simply reflect the time elapsed since they shared the last common ancestor.

Evolutionists debate whether the neutrality theory is valid. Tests of the molecular clock hypothesis indicate that the variations in the rates of molecular evolution are substantially larger than would be expected according to the neutrality theory. Other tests have revealed substantial discrepancies between the amount of genetic polymorphism found in populations of a given species and the amount predicted by the theory. But defenders of the theory argue that these discrepancies can be assimilated by modifying the theory somewhat – by assuming, for example, that alleles are not strictly neutral but their differences in selective value are quite small. Be that as it may, the neutrality theory provides a "null hypothesis", or point of departure, for measuring molecular evolution.

PART 3

GENETICS AND HUMAN HEALTH

6

HUMAN GENETICS

Introduction

The study of human heredity occupies a central position in genetics. Much of this interest stems from a basic desire to know who humans are and why they are as they are. At a more practical level, an understanding of human heredity is of critical importance in the prediction, diagnosis, and treatment of diseases that have a genetic component. The quest to determine the genetic basis of human health has given rise to the field of medical genetics. In general, medicine has given focus and purpose to human genetics, so that the terms medical genetics and human genetics are often considered synonymous. This chapter, however, focuses on the role that genetics has in regulating the human systems in non-pathological situations (i.e. in immune regulation and in blood formation). Studies of twins have also provided much useful information about the effect the environment has on the expression of our genes.

Immunogenetics

Immunity is the ability of an individual to recognize the "self" molecules that make up one's own body and to distinguish them from such "non-self" molecules as those found in infectious microorganisms and toxins. This process has a prominent genetic component. Knowledge of the genetic and molecular basis of the mammalian immune system has increased in parallel with the explosive advances made in molecular genetics.

There are two major components of the immune system, both originating from the same precursor cells known as stem cells. The bursa component provides B lymphocytes, a class of white blood cells that, when appropriately stimulated, differentiate into plasma cells. These latter cells produce circulating soluble proteins called antibodies or immunoglobulins. Antibodies are produced in response to substances called antigens, most of which are foreign proteins or polysaccharides. An antibody molecule can recognize a specific antigen, combine with it, and initiate its destruction. This so-called humoral immunity is accomplished through a complicated series of interactions with other molecules and cells; some of these interactions are mediated by another group of lymphocytes, the T lymphocytes, which are derived from the thymus gland. Once a B lymphocyte has been exposed to a specific antigen, it "remembers" the contact so that future exposure will cause an accelerated and magnified immune reaction. This is a manifestation of what has been called immunological memory.

The thymus component of the immune system centres on the thymus-derived T lymphocytes. In addition to regulating the B cells in producing humoral immunity, the T cells also directly attack cells that display foreign antigens. This process, called cellular immunity, is of great importance in protecting the body against a variety of viruses as well as cancer cells. Cellular immunity is also the chief cause of the rejection of organ transplants. The T lymphocytes provide a complex network consisting of a series of helper cells (which are antigen specific), amplifier cells, suppressor cells, and cytotoxic (killer) cells, all of which are important in immune regulation.

Genetics of Antibody Formation

One of the central problems in understanding the genetics of the immune system has been in explaining the genetic regulation of antibody production. Immunobiologists have demonstrated that the system can produce well over a million specific antibodies, each corresponding to a particular antigen. It would be difficult to envisage that each antibody is encoded by a separate gene – such an arrangement would require a disproportionate share of the entire human genome. Recombinant DNA analysis has illuminated the mechanisms by which a limited number of immunoglobulin genes can encode this vast number of antibodies.

Each antibody molecule consists of several different polypeptide chains – the light chains (L) and the longer heavy chains (H). The latter determine to which of five different classes (IgM, IgG, IgA, IgD, or IgE) an immunoglobulin belongs. Both the L and H chains are unique among proteins in that they contain constant and variable parts. The constant parts have relatively identical amino acid sequences in any given antibody. The variable parts, on the other hand, have different amino acid sequences in each antibody molecule. It is the variable parts, then, that determine the specificity of the antibody.

Recombinant DNA studies of immunoglobulin genes in mice have revealed that the light-chain genes are encoded in four separate parts in germline DNA: a leader segment (L), a variable segment (V), a joining segment (J), and a constant segment (C). These segments are widely separated in the DNA of an embryonic cell, but in a mature B lymphocyte they are found in relative proximity (albeit separated by introns). The mouse has more than 200 light-chain variable region genes, only one of which will be incorporated into the proximal sequence that codes for the antibody production in a given B lymphocyte. Antibody diversity is greatly enhanced by this system, as the V and J segments rearrange and assort randomly in each B-lymphocyte precursor cell. The mechanisms by which this DNA rearrangement takes place are not clear, but transposons are undoubtedly involved. Similar combinatorial processes take place in the genes that code for the heavy chains; furthermore, both the light-chain and heavy-chain genes can undergo somatic mutations to create new antibody-coding sequences. The net effect of these combinatorial and mutational processes enables the coding of millions of specific antibody molecules from a limited number of genes. It should be stressed, however, that each B lymphocyte can produce only one antibody. It is the B lymphocyte population as a whole that produces the tremendous variety of antibodies in humans and other mammals.

Plasma cell tumours (myelomas) have made it possible to study individual antibodies since these tumours, which are descendants of a single plasma cell, produce one antibody in abundance. Another method of obtaining large amounts of a specific antibody is by fusing a B lymphocyte with a rapidly growing cancer cell. The resultant hybrid cell, known as a hybridoma, multiplies rapidly in culture. Since the antibodies obtained from hybridomas are produced by clones derived from a single lymphocyte, they are called monoclonal antibodies.

Genetics of Cellular Immunity

As has been stated, cellular immunity is mediated by T lymphocytes that can recognize infected body cells, cancer cells, and the cells of a foreign transplant. The control of cellular immune reactions is provided by a linked group of genes, known as the major histocompatibility complex (MHC). These genes code for the major histocompatibility antigens, which are found on the surface of almost all nucleated somatic cells. The major histocompatibility antigens were first discovered on the leucocytes (white blood cells) and are, therefore, usually referred to as the HLA (human leucocyte group A) antigens.

The advent of the transplantation of human organs in the 1950s made the question of tissue compatibility between donor and recipient of vital importance, and it was in this context that the HLA antigens and the MHC were elucidated. Investigators found that the MHC resides on the short arm of chromosome 6, on four closely associated sites designated HLA-A, HLA-B, HLA-C, and HLA-D. Each locus is highly polymorphic, i.e. each is represented by a great many alleles within the human gene pool. These alleles, like those of the ABO blood group system, are expressed in co-dominant fashion. Because of the large number of alleles at each HLA locus, there is an extremely low probability of any two individuals (other than siblings) having identical HLA genotypes. (Since a person inherits one chromosome 6 from each parent, siblings have a 25 per cent probability of having received the same paternal and maternal chromosomes 6 and thus of being HLA matched.)

Although HLA antigens are largely responsible for the rejection of organ transplants, it is obvious that the MHC did not evolve to prevent the transfer of organs from one person to another. Indeed, information obtained from the histocompatibility complex in the mouse (which is very similar in its genetic organization to that of the human) suggests that a primary function of the HLA antigens is to regulate the number of specific cytotoxic T killer cells, which have the ability to destroy virus-infected cells and cancer cells.

Genetics of Human Blood

More is known about the genetics of the blood than about any other human tissue. One reason for this is that blood samples can be easily secured and subjected to biochemical analysis without harm or major discomfort to the person being tested. Perhaps a more cogent reason is that many chemical properties of human blood display relatively simple patterns of inheritance.

Blood Types

Certain chemical substances within the red blood cells (such as the ABO and MN substances) may serve as antigens. When cells that contain specific antigens are introduced into the body of an experimental animal such as a rabbit, the animal responds by producing antibodies in its own blood.

In addition to the ABO and MN systems, geneticists have identified about 14 blood-type gene systems associated with other chromosomal locations. The best known of these is the rhesus (Rh) system. The Rh antigens are of particular importance in human medicine. Curiously, however, their existence was discovered in monkeys. When blood from the rhesus monkey (hence the designation Rh) is injected into rabbits, the rabbits produce so-called Rh antibodies that will agglutinate not only the red blood cells of the monkey but the cells of a large proportion of human beings as well. Some people (Rh-negative individuals), however, lack the Rh antigen; the proportion of such persons varies from one human population to another. Akin to data concerning the ABO system, the evidence for Rh genes indicates that only a single chromosome locus (called *r*) is involved and is located on chromosome 1. At least 35 Rh alleles are known for the *r* location; basically the Rh-negative condition is recessive.

A medical problem may arise when a woman who is Rh-negative carries a fetus that is Rh-positive. The first such child may have no difficulty, but later similar pregnancies may produce severely anaemic newborn infants. Exposure to the red blood cells of the first Rh-positive fetus appears to immunize the Rh-negative mother; that is, she develops antibodies that may produce permanent (sometimes fatal) brain damage in any subsequent Rh-positive fetus. Damage arises from the scarcity of oxygen reaching the fetal brain because of the severe destruction of red blood cells. Measures are available for avoiding the severe effects of Rh incompatibility by transfusions to the fetus within the uterus; however, genetic counselling before conception is helpful so that the mother can receive Rh immunoglobulin immediately after her first and any subsequent pregnancies involving an Rh-positive fetus. This immunoglobulin effectively destroys the fetal red blood cells before the mother's immune system is stimulated. The mother thus avoids becoming actively immunized against the Rh antigen and will not produce antibodies that could attack the red blood cells of a future Rh-positive fetus.

Serum Proteins

Human serum, the fluid portion of the blood that remains after clotting, contains various proteins that have been shown to be under genetic control. Study of genetic influences has flourished since the development of precise methods for separating and identifying serum proteins. These move at different rates under the impetus of an electrical field (electrophoresis), as do proteins from many other sources (e.g. muscle or nerve). Since the composition of a protein is specified by the structure of its corresponding gene, biochemical studies based on electrophoresis permit direct study of tissue substances that are only a metabolic step or two away from the genes themselves.

Electrophoretic studies have revealed that at least one-third of the human serum proteins occur in variant forms. Many of the serum proteins are polymorphic, occurring as two or more variants with a frequency of not less than 1 per cent each in a population. Patterns of polymorphic serum protein variants have been used to determine whether twins are identical (as in assessing compatibility for organ transplants) or whether two individuals are related (as in resolving paternity suits). Whether or not the different forms have a selective advantage is not generally known.

Much attention in the genetics of substances in the blood has been centred on serum proteins called haptoglobins, transferrins (which transport iron), and gamma globulins (a number of which are known to immunize against infectious diseases). Haptoglobins appear to relate to two common alleles at a single chromosome locus; the mode of inheritance of the other two seems more complicated, about 18 kinds of transferrins having been described. Like blood-cell antigen genes, serum-protein genes are distributed worldwide in the human population in a way that permits their use in tracing the origin and migration of different groups of people.

Haemoglobin

Hundreds of variants of haemoglobin have been identified by electrophoresis, but relatively few are frequent enough to be called polymorphisms. Of the polymorphisms, the alleles for sickle-cell and thalassaemia haemoglobins produce serious disease in homozygotes, whereas others (haemoglobins C, D, and E) do not. The sickle-cell polymorphism confers a selective advantage on the heterozygote living in a malarial environment; the thalassaemia polymorphism provides a similar advantage.

Genetics of Smell

In 1991 American geneticists Linda Buck and Richard Axel jointly published a landmark scientific paper, based on research they had conducted with laboratory rats, that detailed their discovery of the family of 1,000 genes that encode, or produce, an equivalent number of olfactory receptors. These receptors are proteins responsible for detecting the odorant molecules in the air and are located on olfactory receptor cells, which are clustered within a small area in the back of the nasal cavity. The two scientists clarified how the olfactory system functions by showing that each receptor cell has only one type of odour receptor, which is specialized to recognize a few odours. After odorant molecules bind to receptors, the receptor cells send electrical signals to the olfactory bulb in the brain. The brain combines information from several types of receptors in specific patterns, which are experienced as distinct odours.

Axel and Buck later determined that most of the details they uncovered about the sense of smell are virtually identical in rats, humans, and other animals, although they discovered that humans have only about 350 types of working olfactory receptors, about one-third the number in rats. Nevertheless, the genes that encode olfactory receptors in humans account for about 3 per cent of all human genes. The work helped boost scientific interest in the possible existence of human pheromones, odorant molecules known to trigger sexual activity and certain other behaviour in many animals, and Buck's laboratory continued to investigate how odour perceptions are translated into emotional responses and instinctive behaviour.

Influence of the Environment

Gene expression is modified by the environment. A good example is the recessively inherited disease called galactosaemia, in which the enzyme necessary for the metabolism of galactose – a component of milk sugar – is defective. The sole source of galactose in the infant's diet is milk, which in this instance is toxic. The treatment of this most serious disease in the neonate is to remove all natural forms of milk from the diet (environmental manipulation) and to substitute a synthetic milk lacking galactose. The infant will then develop normally but will never be able to tolerate foods containing lactose. If milk were not a major part of the infant's diet, however, the mutant gene would never be able to express itself, and galactosaemia would be unknown.

Another way of saying this is that no trait can exist or become actual without an environmental contribution. Thus, the old question of which is more important, heredity or environment, is without meaning. Both nature (heredity) and nurture (environment) are always important for every human attribute.

But this is not to say that the separate contributions of heredity and environment are equivalent for each characteristic. Dark pigmentation of the iris of the eye, for example, is under hereditary control in that one or more genes specify the synthesis and deposition in the iris of the pigment (melanin). This is one character that is relatively independent of such environmental factors as diet or climate; thus, individual differences in eye colour tend to be largely attributable to hereditary factors rather than to ordinary environmental change.

On the other hand, it is unwarranted to assume that other traits (such as height, weight, or intelligence) are as little affected by environment as is eye colour. It is very easy to gather information that tall parents tend, on average, to have tall children (and that short parents tend to produce short children), properly indicating a hereditary contribution to height. Nevertheless, it is equally manifest that growth can be stunted in the environmental absence of adequate nutrition. The dilemma arises that only the combined, final result of this nature–nurture interaction can be directly observed. There is no accurate way (in the case of a single individual) to gauge the separate contributions of heredity and environment to such a characteristic as height. An inferential way out of this dilemma is provided by studies of twins.

Fraternal Twins

Usually a fertile human female produces a single egg about once a month. Should fertilization occur (a zygote is formed), growth of the individual child normally proceeds after the fertilized egg has become implanted in the wall of the uterus (womb). In the unusual circumstance that two unfertilized eggs are simultaneously released by the ovaries, each egg may be fertilized by a different sperm cell at about the same time, become implanted, and grow, to result in the birth of twins.

Twins formed from separate eggs and different sperm cells can be of the same or of either sex. No matter what their sex, they are designated as fraternal twins. This terminology is used to emphasize that fraternal twins are genetically no more alike than are siblings (brothers or sisters) born years apart. They differ from ordinary siblings only in having grown side by side in the womb and in having been born at approximately the same time.

Identical Twins

In a major non-fraternal type of twinning, only one egg is fertilized; but during the cleavage of this single zygote into two cells, the resulting pair somehow become separated. Each of the two cells may implant in the uterus separately and grow into a complete, whole individual. In laboratory studies with the zygotes of many animal species, it has been found that in the two-cell stage (and later) a portion of the embryo, if separated under the microscope by the experimenter, may develop into a perfect, whole individual. Such splitting occurs spontaneously at the four-cell stage in some organisms (e.g. the armadillo) and has been accomplished experimentally with the embryos of salamanders, among others.

The net result of splitting at an early embryonic stage may be to produce so-called identical twins. Since such twins derive from the same fertilized egg, the hereditary material from which they originate is absolutely identical in every way, down to the last gene locus. While developmental and genetic differences between one "identical" twin and another still may arise through a number of processes (e.g. mutation), these twins are always of the same sex. They are often breathtakingly similar in appearance, frequently down to very fine anatomic and biochemical details (although their fingerprints are differentiable).

Diagnosis of Twin Types

Since the initial event in the mother's body (either splitting of a single egg or two separate fertilizations) is not observed directly, inferential means are employed for diagnosing a set of twins as fraternal or identical. The birth of fraternal twins is frequently characterized by the passage of two separate afterbirths. In many instances, identical twins are followed by only a single afterbirth, but exceptions to this phenomenon are so common that this is not a reliable method of diagnosis.

The most trustworthy method for inferring twin type is based on the determination of genetic similarity. By selecting those traits that display the least variation attributable to environmental influences (such as eye colour and blood types), it is feasible, if enough separate chromosome loci are considered, to make the diagnosis of twin type with high confidence. HLA antigens, which, as stated above, are very polymorphic, have become most useful in this regard.

Inferences from Twin Studies

Metric (quantitative) traits

By measuring the heights of a large number of ordinary siblings (brothers and sisters) and of twin pairs, it may be shown that the average difference between identical twins is less than half the difference for all other siblings. Any average differences between groups of identical twins are attributable with considerable confidence to the environment. Thus, since the sample of identical twins who were reared apart (in different homes) differed little in height from identicals who were raised together, it appears that environmental–genetic influences on that trait tended to be similar for both groups.

Yet the data for like-sexed fraternal twins reveal a much greater average difference in height (about the same as that found for ordinary siblings reared in the same home at different ages). Apparently the fraternal twins were more dissimilar than identicals (even though reared together) because the fraternals differed more among themselves in genotype. This emphasizes the great genetic similarity among identicals. Such studies can be particularly enlightening when the effects of individual genes are obscured or distorted by the influence of environmental factors on quantitative (measurable) traits (e.g. height, weight, and intelligence).

Any trait that can be objectively measured among identical and fraternal twins can be scrutinized for the particular combination of hereditary and environmental influences that impinge upon it. The effect of environment on identical twins reared apart is suggested by their relatively great average difference in body weight as compared with identical twins reared together. Weight appears to be more strongly modified by environmental variables than is height.

Studies of comparable characteristics among farm animals and plants suggest that such quantitative human traits as height and weight are affected by allelic differences at a number of chromosome locations, meaning that they are not simply affected by genes at a single locus. Investigation of these gene systems with multiple locations (polygenic systems) is carried out largely through selective-breeding experiments among large groups of plants and lower animals. Human beings select their mates in a much freer fashion, of course, and polygenic studies among people are thus severely limited.

Intelligence is a very complex human trait, the genetics of which has been a subject of controversy for some time. Much of the controversy arises from the fact that intelligence is so difficult to define. Information has been based almost entirely on scores on standardized IQ tests constructed by psychologists; in general such tests do not take into account cultural, environmental, and educational differences. As a result, the working definition of intelligence has been "the general factor common to a large number of diverse cognitive (IQ) tests". Even roughly measured as IQ, intelligence shows a strong contribution from the environment. Fraternal twins, however, show relatively great dissimilarity in IQ, suggesting an important contribution from heredity as well. In fact, it has been estimated that, on average, between 60 and 80 per cent of the variance in IQ test scores could be genetic. It is important to note that intelligence is polygenically inherited and that it has the highest degree of assortative mating of any trait; in other words, people tend to mate with people having similar IQs. Moreover, twin studies involving psychological traits should be viewed with caution; for example, since identical twins tend to be singled out for special attention, their environment should not be considered equivalent even to that of other children raised in their own family.

Since the time of British scientist Sir Francis Galton, who conducted pioneering studies of human intelligence in the 19th century, generalizations have been repeatedly made about racial differences in intelligence, with claims of genetic superiority of some races over others. These generalizations fail to recognize that races are composed of individuals, each of whom has a unique genotype made up of genes shared with other humans, and that the sources of intraracial variation are more numerous than those producing interracial differences.

Other traits

For traits of a more qualitative (all-or-none) nature, the twin method can also be used in efforts to assess the degree of hereditary contribution. Such investigations are based on an examination of cases in which at least one member of the twin pair shows the trait. It was found in one study, for example, that in about 80 per cent of all identical twin pairs in which one twin shows symptoms of the psychiatric disorder called schizophrenia, the other member of the pair also shows the symptoms (that is, the two are concordant for the schizophrenic trait). In the remaining 20 per cent, the twins are discordant (that is, one lacks the trait). Since identical twins often have similar environments, this information by itself does not distinguish between the effects of heredity and environment. When pairs of like-sexed fraternal twins reared together are studied, however, the degree of concordance for schizophrenia is very much lower – only about 15 per cent.

Schizophrenia thus clearly develops much more easily in some genotypes than in others; this indicates a strong hereditary predisposition to the development of the trait. Schizophrenia also serves as a good example of the influence of environmental factors, since concordance for the condition does not appear in 100 per cent of identical twins.

Studies of concordance and discordance between identical and fraternal twins have been carried out for many other human characteristics. It has, for example, been known for many years that tuberculosis is a bacterial infection of environmental origin. Yet identical twins raised in the same home show concordance for the disease far more often than do fraternal twins. This finding seems to be explained by the high degree of genetic similarity among the identical twins. The tuberculosis germ is not inherited, but heredity does seem to make one more (or less) susceptible to this particular infection.

Thus, the genes of one individual may provide the chemical basis for susceptibility to a disease, while the genes of another may fail to do so.

Indeed, there seem to be genetic differences among disease germs themselves that result in differences in their virulence. Thus, whether a genetically susceptible person actually develops a disease also depends in part on the heredity of the particular strain of bacteria or virus with which he or she must cope. Consequently, unless environmental factors such as these are adequately evaluated, the conclusions drawn from susceptibility studies can be unfortunately misleading.

The above discussion should help to make clear the limits of genetic determinism. The expression of the genotype can always be modified by the environment. It can be argued that all human illnesses have a genetic component and that the basis of all medical therapy is environmental modification. Specifically, this is the hope for the management of genetic diseases. The more that can be learned about the basic molecular and cellular dysfunctions associated with such diseases, the more amenable they will be to environmental manipulation.

7

HUMAN GENETIC DISEASE

Introduction

With the increasing ability to control infectious and nutritional diseases in developed countries, there has come the realization that genetic diseases are a major cause of disability, death, and human tragedy. Rare, indeed, is the family that is entirely free of any known genetic disorder. Many thousands of different genetic disorders with defined clinical symptoms have been identified. Of the 3 to 6 per cent of newborns with a recognized birth defect, at least half involve a predominantly genetic contribution. Furthermore, genetic defects are the major known cause of pregnancy loss in developed countries, and almost half of all spontaneous abortions (miscarriages) involve a chromosomally abnormal fetus. About 30 per cent of all postnatal infant mortality in developed countries is due to genetic disease; 30 per cent of paediatric and 10 per cent of adult hospital admissions can be traced to a predominantly genetic cause. Finally, medical investigators estimate that genetic defects – however minor –

are present in at least 10 per cent of all adults. Thus, these are not rare events.

A congenital defect is any biochemical, functional, or structural abnormality that originates before or shortly after birth. It must be emphasized that birth defects do not all have the same basis, and it is even possible for apparently identical defects in different individuals to reflect different underlying causes. Though the genetic and biochemical bases for most recognized defects are still uncertain, it is evident that many of these disorders result from a combination of genetic and environmental factors.

This chapter surveys the main categories of genetic disease, focusing on the types of genetic mutations that give rise to them, the risks associated with exposure to certain environmental agents, and the course of managing genetic disease through counselling, diagnosis, and treatment.

Classes of Genetic Disease

Most human genetic defects can be categorized as resulting from either chromosomal, single-gene Mendelian, single-gene non-Mendelian, or multifactorial causes. Each of these categories is discussed briefly below.

Diseases Caused by Chromosomal Aberrations

About 1 out of 150 live newborns has a detectable chromosomal abnormality. Yet even this high incidence represents only a small fraction of chromosome mutations, since the vast majority are lethal and result in prenatal death or stillbirth. Indeed, 50 per cent of all miscarriages in the first three months of pregnancy (the first trimester) and 20 per cent of all miscarriages in the second three months (the second trimester) are estimated to involve a chromosomally abnormal fetus.

Chromosome disorders can be grouped into three principal categories: (1) those that involve numerical abnormalities of the autosomes, (2) those that involve structural abnormalities of the autosomes, and (3) those that involve the sex chromosomes. Autosomes are the 22 sets of chromosomes found in all normal human cells. They are referred to numerically (e.g. chromosome 1, chromosome 2) according to a traditional sort order based on size, shape, and other properties. Sex chromosomes make up the 23rd pair of chromosomes in all normal human cells and come in two forms, known as X and Y. In humans and many other animals, it is the constitution of sex chromosomes that determines the sex of the individual, such that XX results in a female and XY results in a male.

Numerical abnormalities

Numerical abnormalities, involving either the autosomes or sex chromosomes, are believed generally to result from meiotic non-disjunction – that is, the unequal division of chromosomes between daughter cells – that can occur during either maternal or paternal gamete formation. Meiotic non-disjunction leads to eggs or sperm with additional or missing chromosomes. Although the biochemical basis of numerical chromosome abnormalities remains unknown, maternal age clearly has an effect. Older women are at significantly increased risk of conceiving and giving birth to a chromosomally abnormal child. The risk increases with age, especially after age 35, so that a pregnant woman aged 45 or older has between a 1 in 30 and 1 in 100 chance that her child will have trisomy 21 (Down syndrome), while the risk is only 1 in 400 for a 35-year-old woman and less than 1 in 1,000 for a woman under the age of 30. There is no clear effect of paternal age on numerical chromosome abnormalities.

Although Down syndrome is probably the best-known and most commonly observed of the autosomal trisomies, being found in about 1 out of 740 live births, both trisomy 13 and trisomy 18 are also seen in the population, although they are very rare (1 out of 10,000 live births and 1 out of 6,000 live births, respectively). The vast majority of conceptions involving trisomy for any of these three autosomes are lost to miscarriage, as are all conceptions involving trisomy for any of the other autosomes. Similarly, monosomy for any of the autosomes is lethal in utero and therefore is not seen in the population.

Because numerical chromosomal abnormalities generally result from independent meiotic events, parents who have one pregnancy with a numerical chromosomal abnormality generally do not have a markedly increased risk above the general population of repeating the experience. Nonetheless, a small increased risk is generally cited for these couples to account for unusual situations, such as chromosomal translocations or gonadal mosaicism, described below.

Structural abnormalities

Structural abnormalities of the autosomes are even more common in the population than are numerical abnormalities. They include translocations of large pieces of chromosomes, as well as smaller deletions, insertions, or rearrangements. Indeed, about 5 per cent of all cases of Down syndrome result not from classic trisomy 21 but from the presence of excess chromosome 21 material attached to the end of another chromosome as the result of a translocation event.

Balanced, structural chromosomal abnormalities may be compatible with a normal phenotype, although unbalanced chromosome structural abnormalities can be every bit as devastating as numerical abnormalities. Furthermore, because many structural defects are inherited from a parent who is a balanced carrier, couples who have one pregnancy with a structural chromosomal abnormality generally are at significantly increased risk above the general population to repeat the experience. Clearly, the likelihood of a recurrence would depend on whether a balanced form of the structural defect occurs in one of the parents.

Even a small deletion or addition of autosomal material – too small to be seen by normal karyotyping methods – can produce serious malformations and intellectual disability. One example is *cri-du-chat* (French: "cry of the cat") syndrome, which is associated with the loss of a small segment of the short arm of chromosome 5. Newborns with this disorder have a "mewing" cry, like that of a cat, and intellectual disability, which is usually severe.

Abnormalities of the sex chromosomes

About 1 in 400 male and 1 in 650 female live births demonstrate some form of sex chromosome abnormality, although the symptoms of these conditions are generally much less severe than are those associated with autosomal abnormalities. Turner syndrome is a condition of females who, in the classic form, carry only a single X chromosome, giving them a karyotype of 45,X (the normal is 46, XX for a human female and 46, XY for a male). Turner syndrome is characterized by a collection of symptoms, including short stature, webbed neck, and incomplete or absent development of secondary sex characteristics, leading to infertility. Although Turner syndrome is seen in about 1 in 2,500 to 1 in 5,000 female live births, the 45,X karyotype accounts for 10 to 20 per cent of the chromosomal abnormalities seen in spontaneously aborted fetuses, demonstrating that almost all 45,X conceptions are lost to miscarriage. Indeed, the majority of liveborn females with Turner syndrome are diagnosed as mosaics, meaning that some proportion of their cells are 45,X while the rest are either 46,XX or 46,XY. The degree of clinical severity generally correlates inversely with the degree of mosaicism, so that females with a higher proportion of normal cells will tend to have a milder clinical outcome.

In contrast to Turner syndrome, which results from the absence of a sex chromosome, three alternative conditions result from the presence of an extra sex chromosome: Klinefelter syndrome, trisomy X, and 47,XYY syndrome. These conditions, each of which occurs in about 1 in 1,000 live births, are clinically mild, perhaps reflecting the fact that the Y chromosome carries relatively few genes, and, although the X chromosome is gene-rich, most of these genes become transcriptionally silent in all but one X chromosome in each somatic cell (i.e. all cells except eggs and sperm) via a process called X inactivation.

The phenomenon of X inactivation prevents a female who carries two copies of the X chromosome in every cell from expressing twice the amount of gene products encoded exclusively on the X chromosome, in comparison with males, who carry a single X. In brief, at some point in early development one X chromosome in each somatic cell of a female embryo undergoes chemical modification and is inactivated so that gene expression no longer occurs from that template. This process is apparently random in most embryonic tissues, so that roughly half of the cells in each somatic tissue will inactivate the maternal X while the other half will inactivate the paternal X. Cells destined to give rise to eggs do not undergo X inactivation, and cells of the extra-embryonic tissues preferentially inactivate the paternal X, although the rationale for this preference is unclear. The inactivated X chromosome typically replicates later than other chromosomes, and it physically condenses to form a Barr body, a small structure found at the rim of the nucleus in female somatic cells between divisions. The discovery of X inactivation is generally attributed to British geneticist Mary Lyon, and it is therefore often called "lyonization".

The result of X inactivation is that all normal females are mosaics with regard to this chromosome, meaning that they are composed of some cells that express genes only from the maternal X chromosome and others that express genes only from the paternal X chromosome. Although the process is apparently random, not every female has an exact 1 : 1 ratio of maternal to paternal X inactivation. Indeed, studies suggest that ratios of X inactivation can vary. Furthermore, not all genes on the X chromosome are inactivated; a small number escape modification and remain actively expressed from both X chromosomes in the cell. Although this class of genes has not yet been fully characterized, aberrant expression of these genes has been raised as one possible explanation for the phenotypic abnormalities experienced by individuals with too few or too many X chromosomes.

Klinefelter syndrome (47,XXY) occurs in males and is associated with increased stature and infertility. Gynaecomastia (i.e. partial breast development in a male) is sometimes also seen. Males with Klinefelter syndrome, like normal females, inactivate one of their two X chromosomes in each cell, perhaps explaining, at least in part, the relatively mild clinical outcome.

Trisomy X (47,XXX) is seen in females and is generally also considered clinically benign, although menstrual irregularities or sterility have been noted in some cases. Females with trisomy X inactivate two of the three X chromosomes in each of their cells, again perhaps explaining the clinically benign outcome.

47,XYY syndrome also occurs in males and is associated with tall stature but few, if any, other clinical manifestations. There is some evidence of mild learning disability associated with each of the sex chromosome trisomies, although there is no evidence of intellectual disability (a more severe, cognitive deficiency) in these persons.

Persons with karyotypes of 48,XXXY or 49,XXXXY have been reported but are extremely rare. These individuals show clinical outcomes similar to those seen in males with Klinefelter syndrome but with slightly increased severity. In these persons the "n - 1 rule" for X inactivation still holds, so that all but one of the X chromosomes present in each somatic cell is inactivated.

Diseases Associated with Single-gene Mendelian Inheritance

The term Mendelian is often used to denote patterns of genetic inheritance similar to those described for traits in the garden pea by Gregor Mendel in the 1860s. Disorders associated with single-gene Mendelian inheritance are typically categorized as autosomal dominant, autosomal recessive, or sex-linked. Each category is described briefly in this section. For a full explanation of Mendelian genetics and of the concepts of dominance and recessiveness, see Chapter 3.

Autosomal dominant inheritance

A disease trait that is inherited in an autosomal dominant manner can occur in either sex and can be transmitted by either parent. It manifests itself in the heterozygote (designated Aa), who receives a mutant gene (designated *a*) from one parent and a normal ("wild-type") gene (designated *A*) from the other. In such a case the pedigree (i.e. a pictorial representation of family history) is vertical – that is, the disease passes from one generation to the next. Figure 7.1 illustrates the pedigree for a family with achondroplasia, an autosomal dominant disorder characterized by short-limbed dwarfism that results from a specific mutation in the fibroblast growth factor receptor 3 (*FGFR3*) gene. In pedigrees of this sort, circles refer to females and squares to males; two symbols directly joined at the midpoint represent a mating, and those suspended from a common overhead line represent siblings, with descending birth order from left to right. Solid symbols represent affected individuals, and open symbols represent unaffected individuals. The Roman numerals denote generations, whereas the Arabic numerals identify individuals within each generation. Each person listed in a pedigree may therefore be specified uniquely by a combination of one Roman and one Arabic numeral, such as II-1.

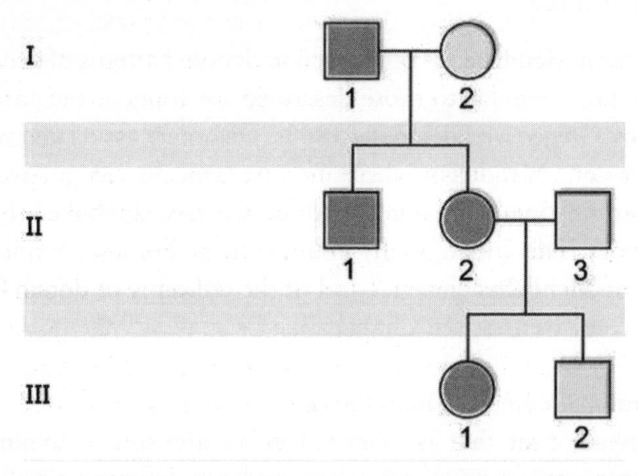

Figure 7.1 Pedigree of a family with a history of achondroplasia, an autosomal dominantly inherited disease. The black symbols signify affected individuals.

An individual who carries one copy of a dominant mutation (Aa) will produce a population of germ cells (i.e. eggs or sperm), in which one half will bear the mutant gene (*A*), and the other will bear the normal gene (*a*). As a result, an affected heterozygote has a 50 per cent chance of passing on the disease gene to each of his or her children. If an individual were to carry two copies of the dominant mutant gene (inherited from both parents), he or she would be homozygous (*AA*). The homozygote for a dominantly inherited abnormal gene may be equally affected with the heterozygote. Alternatively, he or she may be much more seriously affected; indeed, the homozygous condition may be lethal, sometimes even in utero or shortly after birth. Such is the case with achondroplasia, so that a couple with one affected partner and one unaffected partner will typically see half of their children affected, whereas a couple with both partners affected will see two-thirds of their surviving children affected and one-third unaffected, because 1 out of 4 conceptions will produce a homozygous fetus who will die before or shortly after birth.

Although autosomal dominant traits are typically evident in multiple generations of a family, they can also arise from new mutations, so that two unaffected parents, neither of whom carries the mutant gene in their somatic cells, can conceive an affected child. Indeed, for some disorders the new mutation rate is quite high; almost 7 out of 8 children with achondroplasia are born to 2 unaffected parents. Examples of autosomal dominant inheritance are common among human traits and diseases. More than 2,000 of these traits have been clearly identified; a few are listed in Table 7.1.

In many genetic diseases, including those that are autosomal dominant, specific mutations associated with the same disease present in different families may be uniform. This means that every affected individual carries exactly the same molecular defect (allelic homogeneity). Achondroplasia is an example of this; essentially all affected individuals carry exactly the same

Table 7.1 Human disorders attributable to a single dominant gene

Trait	**Conspicuous signs**
Achondroplasia	Dwarfism, large head, short extremities, short fingers and toes
Osteogenesis imperfecta	Bone fragility, deafness
Huntington disease	Involuntary movement, emotional disturbance, dementia
Marfan syndrome	Long, thin extremities and fingers; eye and cardiovascular problems
Neurofibromatosis	Pigmented spots (café au lait) on skin, skin tumours, occasional brain or other internal tumours

mutation. Alternatively, they may be heterogeneous, such that tens or even hundreds of different mutations, all affecting the same gene, may be seen in the affected population (allelic heterogeneity). In some cases even mutations in different genes can lead to the same clinical disorder (genetic heterogeneity).

With regard to the physical manifestations (i.e. the phenotype) of some genetic disorders, a mutant gene may cause many different symptoms and may affect many different organ systems (pleiotropy). For example, along with the short-limbed dwarfism characteristic of achondroplasia, some individuals with this disorder also exhibit a long, narrow trunk, a large head with frontal bossing, and hyperextensibility of most joints, especially the knees. Similarly, for some genetic disorders, clinical severity may vary dramatically, even among affected members in the same family. These variations of phenotypic expression are called variable expressivity, and they are undoubtedly due to the modifying effects of other genes or environmental factors. Although for some disorders, such as achondroplasia, essentially all individuals carrying the mutant gene exhibit the disease phenotype, for other disorders some individuals who carry the mutant gene may express no apparent phenotypic abnormalities at all. Such unaffected individuals are called "non-penetrant", although they can pass on the mutant gene to their offspring, who could be affected.

Autosomal recessive inheritance

Several thousand traits have been related to single genes that are recessive; that is, their effects are masked by normal ("wild-type") dominant alleles and manifest themselves only in individuals homozygous for the mutant gene. A partial list of recessively inherited diseases is given in Table 7.2. For example, sickle cell anaemia, a severe haemoglobin disorder, results only when a mutant gene (*a*) is inherited from both parents. Each of the latter is a carrier, a heterozygote with one normal gene and one mutant gene (*Aa*) who is phenotypically unaffected. The chance of such a couple producing a child with sickle cell anaemia is one out of four for each pregnancy. For couples consisting of one carrier (*Aa*) and one affected individual (*aa*), the chance of their having an affected child is one out of two for each pregnancy.

Table 7.2 Human disorders attributable to a single pair of recessive genes

Trait	Conspicuous signs
Albinism	Lack of pigment in skin, hair, and eyes, with significant visual problems
Tay–Sachs disease	Listlessness, seizures, blindness, death in early childhood
Cystic fibrosis	Chronic lung and intestinal symptoms
Phenylketonuria	Light pigmentation, intellectual disability, seizures
Thalassaemia	Mild or severe anaemia, enlarged spleen and liver, stunted growth, bone deformation
Sickle cell anaemia	Fatigue, shortness of breath, delayed growth, muscle and abdominal pain

Many autosomal recessive traits reflect mutations in key metabolic enzymes and result in a wide variety of disorders classified as inborn errors of metabolism. One of the best-known examples of this class of disorders is phenylketonuria (PKU), which results from mutations in the gene encoding the enzyme phenylalanine hydroxylase (PAH). PAH normally catalyses the conversion of phenylalanine, an amino acid prevalent in dietary proteins and in the artificial sweetener aspartame, to another amino acid called tyrosine. In persons with PKU, dietary phenylalanine either accumulates in the body or some of it is converted to phenylpyruvic acid, a substance that is normally produced only in small quantities. Individuals with PKU tend to excrete large quantities of this acid, along with phenylalanine, in their urine. When infants accumulate high concentrations of phenylpyruvic acid and unconverted phenylalanine in their blood and other tissues, the consequence is intellectual disability. Fortunately, with early detection, strict dietary restriction of phenylalanine, and supplementation of tyrosine, intellectual disability can be prevented.

Since the recessive genes that cause inborn errors of metabolism are individually rare in the gene pool, it is not often that both parents are carriers; hence, the diseases are relatively uncommon. If the parents are related (consanguineous), however, they will be more likely to have inherited the same mutant gene from a common ancestor. For this reason, consanguinity is often more common in the parents of those with rare, recessive inherited diseases. The pedigree of a family in which PKU has occurred is shown in Figure 7.2. This pedigree demonstrates that the affected individuals for recessive diseases are usually siblings in one generation – the pedigree tends to be "horizontal", rather than "vertical" as in dominant inheritance.

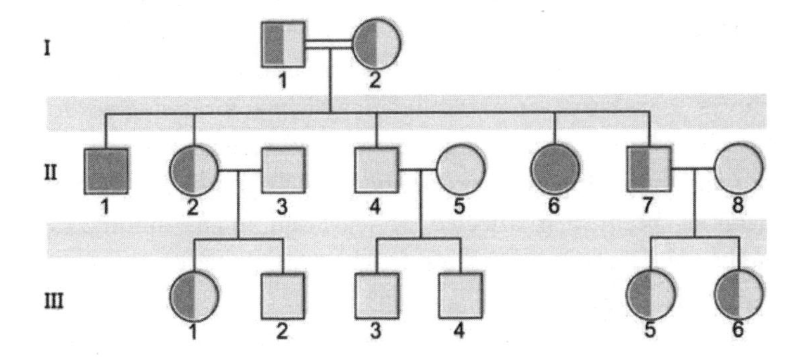

Figure 7.2 Pedigree of a family in which the gene for phenylketonuria is segregating. The half-solid circles and squares represent carriers of phenylketonuria; the solid symbols signify affected individuals. The double line between I-1 and I-2 denotes a consanguineous mating.

Sex-linked inheritance

In humans, there are hundreds of genes located on the X chromosome that have no counterpart on the Y chromosome. The traits governed by these genes thus show sex-linked inheritance. This type of inheritance has certain unique characteristics, which include the following: (1) There is no male-to-male (father-to-son) transmission, since sons will, by definition, inherit the Y rather than the X chromosome. (2) The carrier female (heterozygote) has a 50 per cent chance of passing the mutant gene to each of her children; sons who inherit the mutant gene will be hemizygotes and will manifest the trait, while daughters who receive the mutant gene will be unaffected carriers. (3) Males with the trait will pass the gene on to all of their daughters, who will be carriers. (4) Most sex-linked traits are recessively inherited, so that heterozygous females generally do not display the trait.

Table 7.3 lists some sex-linked conditions and Figure 7.3 shows a pedigree of a family in which a mutant gene for haemophilia A, a sex-linked recessive disease, is segregating. Haemophilia A gained notoriety in early studies of human genetics because it affected at least ten males among the descendants of Queen Victoria, who was a carrier.

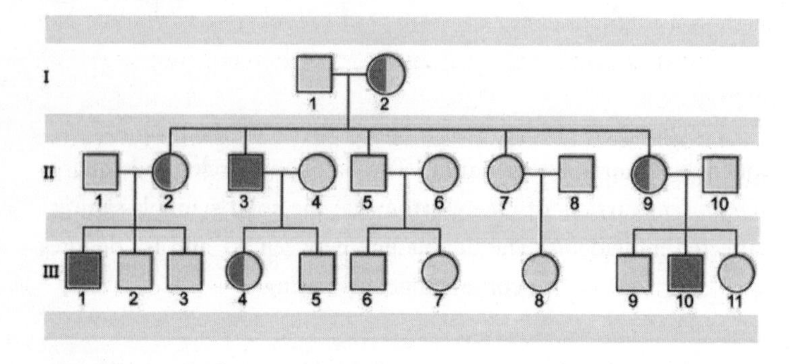

Figure 7.3 Pedigree of a family with a history of haemophilia A, a sex-linked recessively inherited disease. Half-solid circles represent female carriers (heterozygotes) of haemophilia A; the solid squares signify affected males (hemizygotes).

Table 7.3 Human disorders attributable to sex-linked recessive inheritance

Trait	**Conspicuous signs**
Haemophilia	Bleeding tendency with joint involvement
Duchenne muscular dystrophy	Progressive muscle weakness
Lesch–Nyhan syndrome	Cerebral palsy, self-mutilation
Fragile-X syndrome	Intellectual disability, characteristic facies

Haemophilia A, the most widespread form of haemophilia, results from a mutation in the gene encoding a blood protein known as clotting factor VIII. Because of this mutation, affected males cannot produce functional factor VIII, so that their blood fails to clot properly, leading to significant and potentially life-threatening loss of blood after even minor injuries. Bleeding into joints commonly occurs as well and may be crippling. Therapy consists of avoiding trauma and of administering injections of purified factor VIII, which was once isolated from outdated human blood donations but can now be made in large amounts through recombinant DNA technology.

Although heterozygous female carriers of X-linked recessive mutations generally do not exhibit traits characteristic of the disorder, cases of mild or partial phenotypic expression in female carriers have been reported, resulting from non-random X inactivation.

Diseases Associated with Single-gene Non-Mendelian Inheritance

Although disorders resulting from single-gene defects that demonstrate Mendelian inheritance are perhaps better understood, it is clear that a significant number of single-gene diseases also exhibit distinctly non-Mendelian patterns of inheritance. Among these are such disorders that result from triplet repeat expansions within or near specific genes (e.g. Huntington disease and fragile-X syndrome); a collection of neurodegenerative disorders, such as Leber hereditary optic neuropathy, that result from inherited mutations in the mitochondrial DNA; and diseases that result from mutations in imprinted genes (e.g. Angelman syndrome and Prader–Willi syndrome).

Triplet repeat expansions

At least a dozen different disorders are known to result from triplet repeat expansions in the human genome, and these fall into two groups: (1) those that involve a polyglutamine tract within the encoded protein product that becomes longer upon expansion of a triplet repeat (e.g. Huntington disease); and (2) those that have unstable triplet repeats in non-coding portions of the gene that, upon expansion, interfere with appropriate expression of the gene product (e.g. fragile-X syndrome). Both groups of disorders exhibit a distinctive pattern of non-Mendelian inheritance termed anticipation. This means that, once the disorder has initially appeared in a given family, subsequent generations tend to show both increasing frequency and increasing severity of the disorder. This phenotypic anticipation is paralleled by increases in the relevant repeat length as it is passed from one generation to the next, with increasing size leading to increasing instability, until a "full expansion" mutation is achieved, generally several generations after the initial appearance of the disorder in the family. The full expansion mutation is then passed to subsequent generations in a standard Mendelian fashion – for example, autosomal dominant for Huntington disease and sex-linked for fragile-X syndrome.

Mitochondrial DNA mutations

Disorders resulting from mutations in the mitochondrial genome demonstrate an alternative form of non-Mendelian inheritance, termed maternal inheritance, in which the mutation and disorder are passed from mothers – never from fathers – to all of their children. The mutations generally affect the function of the mitochondrion, compromising, among other processes, the production of cellular adenosine triphosphate (ATP). Severity and even penetrance can vary widely for disorders resulting from mutations in the mitochondrial DNA, generally believed to reflect the combined effects of heteroplasmy (i.e. mixed populations of both normal and mutant mitochondrial DNA in a single cell) and other confounding genetic or environmental factors. Close to 50 mitochondrial genetic diseases are currently known.

Imprinted gene mutations

Some genetic disorders result from mutations in imprinted genes. Genetic imprinting involves a sex-specific process of chemical modification to the imprinted genes, so that they are expressed unequally, depending on the sex of the parent of origin. So-called maternally imprinted genes are generally expressed only when inherited from the father, and so-called paternally imprinted genes are generally expressed only when inherited from the mother. The disease gene associated with Prader–Willi syndrome is maternally imprinted, so that although every child inherits two copies of the gene (one maternal, one paternal), only the paternal copy is expressed. If the paternally inherited copy carries a mutation, the child will be left with no functional copies of the gene expressed, and the clinical traits of Prader–Willi syndrome will result. Similarly, the disease gene associated with Angelman syndrome is paternally imprinted, so that although every child inherits two copies of the gene, only the maternal copy is expressed. If the maternally inherited copy carries a mutation, the child again will be left with no functional copies of the gene expressed, and the clinical traits of Angelman syndrome will result. Individuals who carry the mutation but received it from the "wrong" parent can certainly pass it on to their children, although they will not exhibit clinical features of the disorder.

On rare occasions, someone is identified as having an imprinted gene disorder but shows no family history and does not appear to carry any mutation in the expected gene. These cases result from uniparental disomy, meaning that a child has inherited both copies of a given chromosome from the same parent, rather than one from each parent, as is the norm. If any key genes on that chromosome are imprinted in the parent of origin, the child may end up with no expressed copies, and a genetic disorder may result. Other genes may be overexpressed, perhaps also leading to clinical complications. Finally, uniparental disomy can account for very rare instances whereby two parents, only one of whom is a carrier of an autosomal recessive mutation, can nonetheless have an affected child, in the circumstance that the child inherits two mutant copies from the carrier parent.

Diseases Caused by Multifactorial Inheritance

Genetic disorders that are multifactorial in origin represent probably the single largest class of inherited disorders affecting the human population. By definition, these disorders involve the influence of multiple genes, generally acting in concert with environmental factors. Such common conditions as cancer, heart disease, and diabetes are considered multifactorial disorders. Indeed, improvements in the tools used to study this class of disorders have made it possible to assign specific contributing gene loci to a number of common traits and disorders. Identification and characterization of these contributing genetic factors may not only enable improved diagnostic and prognostic indicators but may also identify potential targets for future therapeutic intervention. The wide range of conditions associated with multifactorial inheritance includes alcoholism, Alzheimer disease, cancer, coronary heart disease, diabetes, epilepsy, hypertension, obesity, and schizophrenia.

Because the genetic and environmental factors that underlie multifactorial disorders are often unknown, the risks of recurrence are usually arrived at empirically. In general, it can be said that risks of recurrence are not as great for multifactorial conditions as for single-gene diseases and that the risks vary with the number of relatives affected and the closeness of their relationship. Moreover, close relatives of more severely affected individuals (e.g. those with bilateral cleft lip and cleft palate) are generally at greater risk than those related to persons with a less-severe form of the same condition (e.g. unilateral cleft lip).

Behavioral genetics

Behavioral genetics is the study of the influence of an organism's genetic composition on its behaviour and the interaction of heredity and environment insofar as they affect behaviour. Much behavioral genetic research today focuses on identifying specific genes that affect behavioral dimensions, such as personality and intelligence, and disorders, such as autism, hyperactivity, depression, and schizophrenia.

In contrast to traditional molecular genetic research that focused on rare disorders caused by a single genetic mutation, molecular genetic research on complex behavioral traits and common behavioral disorders is much more difficult because multiple genes are involved and each gene has a relatively small effect. However, some genes identified in animal models have contributed to an improved understanding of complex human behavioral disorders such as reading disability, hyperactivity, autism, and dementia.

Genetics of Cancer

Although at least 90 per cent of all cancers are sporadic, meaning that they do not seem to run in families, nearly 10 per cent of cancers are recognized as familial, and some are actually inherited in an apparently autosomal dominant manner. Cancer may therefore be considered a multifactorial disease, resulting from the combined influence of many genetic factors acting in concert with environmental insults (e.g. ultraviolet radiation, cigarette smoke, and viruses; see below, Genetic Damage from Environmental Agents).

Cancers, both familial and sporadic, generally arise from alterations in one or more of three classes of genes: oncogenes, tumour suppressor genes, and genes whose products participate in genome surveillance – for example, in DNA damage repair. For familial cancers, affected members inherit one mutant copy of a gene that falls into one of the latter two classes. That mutation alone is not sufficient to cause cancer, but it predisposes individuals to the disease because they are now either more sensitive to spontaneous somatic mutations, as in the case of altered tumour suppressor genes, or are more prone to experience mutations, as in the case of impaired DNA repair enzymes. Of course, sporadic cancers can also arise from mutations in these same classes of genes, but because all of the mutations must arise anew (*de novo*) in the individual, as opposed to being inherited, they generally appear only later in life, and they do not run in families.

Retinoblastoma, an aggressive tumour of the eye that typically occurs in childhood, offers perhaps one of the clearest examples of the interplay between inherited and somatic mutations in the genesis of cancer. Current data suggest that 60 to 70 per cent of all cases of retinoblastoma are sporadic, whereas the remainder of cases are inherited. The relevant gene, *RB*, encodes a protein that normally functions as a suppressor of cell cycle progression and is considered a classic tumour suppressor gene. Children who inherit one mutant copy of the *RB* gene are at nearly 100 per cent risk of developing retinoblastoma, because the probability that their one remaining functional *RB* gene will sustain a mutation in at least one retinal cell is nearly assured. In contrast, children who inherit two functional copies of the *RB* gene must experience two mutations at the *RB* locus in the same retinal cell in order to develop retinoblastoma; this is a very rare event. This "two-hit" hypothesis of retinoblastoma formation has provided a foundation upon which most subsequent theories of the genetic origins of familial cancer have been built.

Recent studies of both breast and colorectal cancers have revealed that, like retinoblastoma, these cancers are predominantly sporadic, although a small proportion are clearly familial. Sporadic breast cancer generally appears late in life, whereas the familial forms can present much earlier, often before age 40. For familial breast cancer, inherited mutations in one of two specific genes, *BRCA1* and *BRCA2*, account for about 10 per cent of cases observed. The *BRCA1* and *BRCA2* genes both encode protein products believed to function in the pathways responsible for sensing and responding to DNA damage in cells. While a woman in the general population has about a 10 per cent lifetime risk of developing breast cancer, about 30 to 40 per cent of all women with *BRCA1* or *BRCA2* mutations will develop breast cancer by age 50, and close to 90 per cent will develop the disease by age 80. Women with *BRCA1* mutations are also at increased risk of developing ovarian tumours. As with retinoblastoma, both men and women who carry *BRCA1* or *BRCA2* mutations, whether they are personally affected or not, can pass the mutated gene to their offspring, although carrier daughters are much more likely than carrier sons to develop breast cancer.

Two forms of familial colorectal cancer, hereditary non-polyposis colorectal cancer (HNPCC) and familial adenomatous polyposis (FAP), have also been linked to predisposing mutations in specific genes. Persons with familial HNPCC have inherited mutations in one or more of their DNA mismatch repair genes, predominantly *MSH2* or *MLH1*. Similarly, persons with FAP carry inherited mutations in their *APC* genes, the protein product of which normally functions as a tumour suppressor. For individuals in both categories, the combination of inherited and somatic mutations results in a nearly 100 per cent lifetime risk of developing colorectal cancer.

Although most cancer cases are not familial, all are undoubtedly diseases of the genetic material of somatic cells. Studies of large numbers of both familial and sporadic cancers have led to the conclusion that cancer is a disease of successive mutations. These mutations act in concert to deregulate normal cell growth, provide appropriate blood supply to the growing tumour, and ultimately enable tumour cell movement beyond normal tissue boundaries to achieve metastasis (i.e. the dissemination of cancer cells to other parts of the body).

Many of the agents that cause cancer (e.g. X-rays, certain chemicals) also cause mutations or chromosome abnormalities. For example, a large fraction of sporadic tumours have been found to carry oncogenes, altered forms of normal genes (proto-oncogenes) that have sustained a somatic "gain-of-function" mutation. An oncogene may be carried by a virus, or it can result from a chromosomal rearrangement, as is the case in chronic myelogenous leukaemia, a cancer of the white blood cells characterized by the presence of the so-called Philadelphia chromosome in affected cells. The Philadelphia chromosome arises from a translocation in which one half of the long arm of chromosome 22 becomes attached to the end of the long arm of chromosome 9, creating the dominant oncogene *BCR/ABL* at the junction point. The specific function of the *BCR/ABL* fusion protein is not entirely clear. Another example is Burkitt lymphoma, in which a rearrangement between chromosomes places the *MYC* gene from chromosome 8 under the influence of regulatory sequences that normally control expression of immunoglobulin genes. This deregulation of *MYC*, a protein involved in mediating cell cycle progression, is thought to be one of the major steps in the formation of Burkitt lymphoma.

Genetic Damage from Environmental Agents

Humans are exposed to many agents, both natural and human-made, that can cause genetic damage. Among these agents are viruses; compounds produced by plants, fungi, and bacteria; industrial chemicals; products of combustion; alcohol; ultraviolet and ionizing radiation; and even oxygen. Many of these agents have long been unavoidable, and consequently humans have evolved defences to minimize the damage that these agents cause and have developed ways to repair the damage that cannot be avoided.

Viruses

Viruses survive by injecting their genetic material into living cells with the consequence that the biochemical machinery of the host cell is subverted from serving its own needs to serving the needs of the virus. During this process the viral genome often integrates itself into the genome of the host cell. This integration, or insertion, can occur either in the intergenic regions that make up most of the human genome, or it can occur in the middle of an important regulatory sequence or even in the region coding for a protein – i.e. a gene. In either of the latter two scenarios, the regulation or function of the interrupted gene is lost. If that gene encodes a protein that normally regulates cell division, the result may be unregulated cell growth and division.

Alternatively, some viruses carry dominant oncogenes in their genomes, which can transform an infected cell and start it on the path toward cancer. Furthermore, viruses can cause mutations leading to cancer by killing the infected cell. Indeed, one of the body's defences against viral infection involves recognizing and killing infected cells. The death of cells necessitates their replacement by the division of uninfected cells, and the more cell division that occurs, the greater the likelihood of a mutation arising from the small but finite infidelity in DNA replication. Among the viruses that can cause cancer are Epstein–Barr virus, papilloma viruses, hepatitis B and C viruses, retroviruses such as human immunodeficiency virus (HIV), and herpes virus.

Plants, Fungi, and Bacteria

During the ongoing struggle for survival, organisms have evolved toxic compounds as protection against predators or simply to gain competitive advantage. At the same time, these organisms have evolved mechanisms that make themselves immune to the effects of the toxins that they produce. Plants in particular utilize this strategy since they are rooted in place and cannot escape from predators. One-third of the dry weight of some plants can be accounted for by the toxic compounds that are collectively referred to as alkaloids. *Aspergillus flavus*, a fungus that grows on stored grain and peanuts, produces a powerful carcinogen called aflatoxin that can cause liver cancer. Bacteria produce many proteins that are toxic to the infected host, such as diphtheria toxin. They also produce proteins called bacteriocins that are toxic to other bacteria. Toxins can cause mutations indirectly by causing cell death, which necessitates replacement by cell division, thus enhancing the opportunity for mutation. Cyanobacteria that grow in illuminated surface water produce several carcinogens, such as microcystin, saxitoxin, and cylindrospermopsin, that can also cause liver cancer.

Industrial Chemicals

Tens of thousands of different chemicals are routinely used in the production of plastics, fuels, food additives, and even medicines. Many of these chemicals are mutagens (mutation-inducing), and some have been found to be carcinogenic (cancer-producing) in rats or mice.

Mutagenicity and cancinogenicity

A relatively easy and inexpensive test for mutagenicity, the Ames test, utilizes mutant strains of the bacterium *Salmonella typhimurium* and can be completed in a few days. Testing for carcinogenesis, on the other hand, is very time-consuming and expensive because the test substance must be administered to large numbers of laboratory animals, usually mice, for weeks or months before the tissues can be examined for cancers. For this reason, the number of known mutagens far exceeds the number of known carcinogens.

Furthermore, animal tests for carcinogenesis are not completely predictive of the effects of the test chemical on humans for several reasons. First, the abilities of laboratory animals and humans to metabolize and excrete specific chemicals can differ greatly. In addition, in order to avoid the need to test each chemical at a range of doses, each chemical is usually administered at the maximum tolerated dose. At such high doses, toxicity and cell death occur, necessitating cell replacement by growth and cell division; cell division, in turn, increases the opportunity for mutation and hence for cancer. Alternatively, unusually high doses of a chemical may actually mask the carcinogenic potential of a compound because damaged cells may die rather than survive in mutated form.

Combustion Products

The burning of fossil fuels quite literally powers modern industrial societies. If the combustion of such fuels were complete, the products would be carbon dioxide and water. However, combustion is rarely complete, as is evidenced by the visible smoke issuing from chimneys and from the exhausts of diesel engines. Moreover, in addition to the particulates that humans can see, incomplete combustion produces unseen volatile compounds; some of these, such as the dibenzodioxins, are intensely mutagenic and have been demonstrated to cause cancer in laboratory rodents. Epidemiological data indicate that dioxins are associated with increased risk of a variety of human cancers.

The health consequences of combustion are further increased by impurities in fossil fuels and in the oxygen that supports their burning. For example, coal contains sulfur, mercury, lead, and other elements in addition to carbon. During combustion, sulfur becomes sulfur dioxide and that, in turn, gives rise to sulfurous and sulfuric acids. The mercury in the fuel is emitted as a vapour that is very toxic. Atmospheric nitrogen is oxidized at the high temperature of combustion.

The smoke from a cigarette, drawn directly into the lungs, imparts a large number of particulates, as well as a host of volatile compounds, directly into the airways and alveoli.

Some of the volatile compounds are toxic in their own right and others, such as hydroquinones, slowly oxidize, producing free radicals that can damage DNA. As macrophages in the lungs attempt to engulf and eliminate particulates, they cause the production of mutagenic substances. A large fraction of lung cancers are attributable to cigarette smoking, which is also a risk factor for atherosclerosis, hypertension (high blood pressure), heart attack, and stroke.

Alcohol

Moderate consumption of alcohol (ethanol) is well-tolerated and may even increase lifespan. However, alcohol is a potentially toxic substance and one of its metabolites, acetaldehyde, is a mild mutagen. Hence, it is not surprising that the chronic consumption of alcohol leads to liver cirrhosis and other untoward effects. Over-consumption of alcohol by a pregnant woman can cause her baby to develop fetal alcohol syndrome, which is characterized by low birth weight, intellectual disability, and congenital heart disease.

Ultraviolet Radiation

Due to human activities that result in the release of volatile halocarbon compounds, such as the refrigerant freon and the solvent carbon tetrachloride, the chlorine content of the upper atmosphere is increasing. Chlorine catalyses the decomposition of ozone, which shields the Earth from ultraviolet radiation that is emitted from the sun. The Earth's ozone shield has been progressively depleted, most markedly over the polar regions but also measurably so over the densely populated regions of northern Europe, Australia, and New Zealand. One consequence has been an increase in a variety of skin cancers, including melanoma, in those areas. Steps have been taken to stop the release of halocarbons, but the depletion of the ozone layer will nonetheless persist and may worsen for at least several decades.

The action of ultraviolet light on DNA can lead to covalent linking of adjacent pyrimidine bases. Such pyrimidine dimerization is mutagenic, but this damage can be repaired by an enzyme called photolyase, which utilizes the energy of longer wavelengths of light to cleave the dimers. However, people with a defect in the gene coding for photolyase develop xeroderma pigmentosum, a condition characterized by extreme sensitivity to sunlight. These individuals develop multiple skin cancers on all areas of exposed skin, such as the head, neck, and arms.

Ultraviolet light can also be damaging because of photosensitization, the facilitation of photochemical processes. One way that photosensitizers work is by absorbing a photon and then transferring the energy inherent in that photon to molecular oxygen, thus converting the less-active ground-state molecular oxygen into a very reactive excited state, referred to as singlet oxygen, that can attack a variety of cellular compounds, including DNA. Diseases that have a photosensitizing component include lupus and porphyrias. In addition to photosensitizers that occur naturally in the human body, some foods and medicines (e.g. tetracycline) also act in this way, producing painful inflammation and blistering of the skin after even modest exposure to the sun.

Ionizing Radiation

X-rays and gamma rays are sufficiently energetic to cleave water into hydrogen atoms and hydroxyl radicals and are consequently referred to as ionizing radiation. Ionizing radiation and the products of the cleavage of water are able to damage all biological macromolecules, including DNA, proteins, and polysaccharides, and they have long been recognized as being mutagenic, carcinogenic, and lethal.

People are routinely exposed to natural sources of ionizing radiation, such as cosmic rays, and to radioisotopes, such as carbon-14 and radon. They are also exposed to X-rays and human-made radioisotopes used for diagnostic purposes, and some people have been exposed to radioactive fallout from nuclear weapon tests and reactor accidents. Such exposures would be much more damaging were it not for multiple mechanisms of DNA repair that have evolved to deal with simple errors in replication as well as with damage from naturally occurring sources.

Molecular Oxygen

Molecular oxygen (O_2), although essential for life, must be counted among the environmental toxins and mutagens. Because of its unusual electronic structure, O_2 is most easily reduced not by electron pairs but rather by single electrons added one at a time. As O_2 is converted into water, superoxide (O_2^-), hydrogen peroxide (H_2O_2), and a hydroxyl radical (OH) are produced as intermediates. O_2^- can initiate free-radical oxidation of important metabolites, inactivate certain enzymes, and cause release of iron from specific enzymes. The second intermediate, H_2O_2, is a strong oxidant and can give rise to an even more potent oxidant, namely OH, when it reacts with ferrous iron. Thus, O_2^- and O_2H_2 can collaborate in the formation of the destructive OH and can subsequently lead to DNA damage, mutagenesis, and cell death. Breathing 100 per cent oxygen causes damage to the alveoli, which leads to accumulation of fluid in the lungs. Thus, paradoxically, prolonged exposure to hyperoxia causes death due to lack of oxygen.

Humans have evolved multiple defence systems to counter the toxicity and mutagenicity of O_2. Thus, O_2^- is rapidly converted into O_2, and H_2O_2 is neutralized by a family of enzymes called superoxide dismutases. H_2O_2, in turn, is eliminated by other enzymes called catalases and peroxidases, which convert it into O_2 and water.

A few genetic diseases are known to be related to oxygen radicals or to the enzymes that defend against them. Chronic granulomatous disease (CGD) is caused by a defect in the ability of the phagocytic leukocytes to mount the respiratory burst, part of the body's defence against infection. Upon contacting microorganisms and engulfing them, phagocytes greatly increase their consumption of O_2 (the respiratory burst) while releasing O_2^-, H_2O_2, hypochlorite (HOCl), and other agents that kill the microbe. The reduction of O_2 to O_2^- is caused by a multicomponent enzyme called nicotine adenine dinucleotide phosphate (NADPH) oxidase. A defect in any of the components of this oxidase will lead to the absence of the respiratory burst, giving rise to the constant infections indicative of CGD. Before the discovery and clinical application of antibiotics, people born with CGD died from infection during early childhood.

Another such genetic disease is a familial form of amyotrophic lateral sclerosis (ALS), a motor neuron disease also known as Lou Gehrig disease, which is characterized by late-onset progressive paralysis due to the loss of motor neurons. About 20 per cent of cases of ALS have been shown to result from mutations affecting the enzyme superoxide dismutase. The disease is genetically dominant, so that the mutant enzyme causes the disease even when half of the superoxide dismutase present in cells exists in the normal form. Interestingly, most of the mutant variants retain full catalytic activity.

Genetic Testing

The diagnostic evaluation of a genetic disorder begins with a medical history, a physical examination, and the construction of a family pedigree documenting the diseases and genetic disorders present in the past three generations. This pedigree aids in determining if the problem is sex-linked, dominant, recessive, or not likely to be genetic.

Chemical, radiological, histopathologic, and electrodiagnostic procedures can diagnose basic defects in patients suspected of genetic disease. These include chromosome karyotyping (in which chromosomes are arranged according to a standard classification scheme), enzyme or hormone assays, amino acid chromatography of blood and urine, gene and DNA probes, blood and Rh typing, immunoglobulin determination, electrodiagnostics, and haemoglobin electrophoresis (see also Chapter 10, Management of Genetic Disease).

Prenatal Diagnosis

Prenatal screening is performed if there is a family history of inherited disease, the mother is at an advanced age (see Chapter 7, Numerical abnormalities), a previous child had a chromosomal abnormality, or there is an ethnic indication of risk (Ashkenazic Jews and French Canadians are at increased risk for Tay–Sachs disease; blacks, Arabs, Turks, and others for sickle-cell anaemia; and those of Mediterranean descent for thalassaemia, a type of hereditary anaemia). Parents can be tested before or after conception to determine whether they are carriers.

The most common screening test is an assay of alphafetoprotein (AFP) in the maternal serum. Elevated levels are associated with neural tube defects in the fetus such as spina bifida (defective closure of the spine) and anencephaly (absence of brain tissue). When AFP levels are elevated, a more specific diagnosis is attempted using ultrasonography and amniocentesis to analyse the amniotic fluid for the presence of AFP and acetylcholinesterase. Fetal cells contained in the amniotic fluid also can be cultured and the karyotype (chromosome morphology) determined to identify chromosomal abnormality. Cells for chromosome analysis also can be obtained by chorionic villus sampling, the direct needle aspiration of cells from the chorionic villus (placenta).

Chromosomal Analysis

To obtain a person's karyotype, laboratory technicians grow human cells in tissue culture media. After being stained and sorted, the chromosomes are counted and displayed. The cells are obtained from the blood, skin, or bone marrow or by amniocentesis or chorionic villus sampling, as noted above. The standard karyotype has approximately 400 visible bands, and each band contains up to several hundred genes.

Identification of a chromosomal aberration allows for a more accurate prediction of the risk of its recurrence in future offspring. Karyotyping can be used not only to diagnose aneuploidy, which is responsible for Down, Turner, and Klinefelter syndromes, but also to identify the chromosomal aberrations associated with solid tumours such as nephroblastoma, meningioma, neuroblastoma, retinoblastoma, renal-cell carcinoma, small-cell lung cancer, and certain leukaemias and lymphomas.

DNA Probes

Karyotyping requires a great deal of time and effort and may not always provide conclusive information. It is most useful in identifying very large defects involving hundreds or even thousands of genes.

Techniques such as fluorescent in situ hybridization (FISH) have much higher rates of sensitivity and specificity. FISH also provides results more quickly because no cell culture is required. This technique can detect smaller genetic deletions involving one to five genes. It is also useful in detecting moderate-sized deletions such as those causing Prader–Willi syndrome, which is characterized by a rounded face, low forehead, and intellectual disability.

The analysis of individual genes has been greatly enhanced by the development of recombinant DNA technology. Small DNA fragments can be isolated, and unlimited amounts of cloned material can be produced. Once cloned, the various genes and gene products can be used to study gene function in healthy individuals and those with disease. Recombinant DNA methods can detect any change in DNA, down to a one-base-pair change out of the three billion base pairs in the genome. DNA probes are labelled with radioactive isotopes or fluorescent dyes and used to identify persons who are carriers for autosomal recessive conditions. Disorders that can be detected using this technique include haemophilia A, polycystic kidney disease, sickle-cell disease, Huntington disease, cystic fibrosis, and haemochromatosis.

Biochemical Tests

Biochemical tests primarily detect enzymatic defects such as phenylketonuria, porphyria, and glycogen-storage disease. Although testing of newborns for all these abnormalities is possible, it is not cost-effective, because some are quite rare. Screening requirements for these disorders vary from country to country and depend on whether the disease is sufficiently common, has severe consequences, and can be treated or prevented if diagnosed early; and whether the test can be applied to the entire population at risk.

Ageing

Ageing has many facets. Hence there are a number of theories, each of which may explain one or more aspects of ageing. No single theory explains all of the phenomena of ageing.

Genetic Theories

One theory of ageing assumes that the lifespan of a cell or organism is genetically determined – that the genes of an animal contain a "program" that determines its lifespan, just as eye colour is determined genetically. Although long life is often recognized as a familial characteristic, and short-lived strains of vinegar flies, rats, and mice can be produced by selective breeding, other factors clearly can significantly alter the basic genetic program of ageing.

Another genetic theory of ageing assumes that cell death is the result of "errors" introduced in the formation of key proteins, such as enzymes. Slight differences induced in the transmission of information from the DNA molecules of the chromosomes through RNA molecules to the proper assembly of the large and complex enzyme molecules could result in a molecule of the enzyme that would not "work" properly. These so-called error theories have not yet been firmly established, but studies are in progress.

As cells grow and divide, a small proportion of them undergo mutation. These mutations are then reproduced when the cells again divide. The "somatic mutation" theory of ageing assumes that ageing is due to the gradual accumulation of mutated cells that do not perform normally.

Other theories of ageing focus attention on factors that can influence the expression of a genetically determined "program", in terms of cellular and molecular changes involving the accumulation of waste products, cross-linked molecules, or autoimmune responses. It is possible that ageing in an individual is actually due to a breakdown in the control mechanisms that are required in a complex performance.

Inheritance of Longevity

The inheritance of longevity in animal populations such as vinegar flies and mice is determined by comparing the life tables of numerous inbred populations and some of their hybrids. The longevity of sample populations has been measured for more than 40 inbred strains of mice. Two experiments concur in finding that about 30 per cent of longevity variation in female mice is genetically determined, whereas the heritability in male mice is about 20 per cent. These values are comparable to the heritabilities of some physiological performances, such as lifetime egg or milk production, in domestic animals.

Research indicates that much of the variation in survival time between mouse strains is attributable to differences in inherited susceptibility to specific diseases. An important task of gerontology is to determine the extent of such genetic influences on ageing.

The inheritance of longevity in humans is more difficult to investigate because length of life is influenced by socio-economic and other environmental factors that generate spurious correlations between close relatives. A number of studies have been published, most of them pointing to some degree of heritability with regard to length of life or susceptibility to major diseases, such as cancer and heart disease. Although there is disagreement about the degree of heritability of longevity in humans, the evidence for genetic transmission of susceptibility to coronary heart disease and related diseases is strong. In addition, there is evidence that monozygotic (genetically identical) twins tend to have more similar lifespans than do dizygotic (genetically different, fraternal) twins.

Ageing of Genetic Information Systems

The physical basis of ageing is either the cumulative loss and disorganization of important large molecules (e.g. proteins and nucleic acids) of the body or the accumulation of abnormal products in cells or tissues. A major effort in ageing research has been focused on two objectives: (1) to characterize the molecular disruptions of ageing and to determine if one particular kind is primarily responsible for the observed rate and course of senescence; and (2) to identify the chemical or physical reactions responsible for the age-related degradation of large molecules that have either informational or structural roles.

The working molecules of the body, such as enzymes and contractile proteins, which have short turnover times, are not thought to be sites of primary ageing damage. The DNA molecules of the chromosomes appear to be potential sites of primary damage, because damage to DNA corrupts the genetic message on which the development and function of the organism depend. Damage at a single point in the DNA molecule can be followed by the synthesis of an incorrect protein molecule, which may result in the malfunction or death of the host cell or even of the entire organism. Attention has therefore been given to the somatic mutation hypothesis, which asserts that ageing is the result of an accumulation of mutations in the DNA of somatic (body) cells. Aneuploidy, the occurrence of cells with more or less than the correct (euploid) complement of chromosomes, is especially common. The frequency of aneuploid cells in human females increases from 3 per cent at age 10, to 13 per cent at age 70.

Each DNA molecule consists of two complementary strands coiled around each other in a double helix configuration. Evidence indicates that breaks of the individual strands occur with a higher frequency than was once suspected and that virtually all such breaks are repaired by an enzymatic mechanism that destroys the damaged region and then resynthesizes the excised portion, using the corresponding segment of the complementary strand as a model. The mutation rate for a species is therefore governed more by the competence of its repair mechanism than by the rate at which breaks occur. This may help to explain why the mutation rates of different species are roughly proportional to their generation times and justifies research to determine whether the enzymatic mechanisms involved are accessible to control. It remains to be seen whether a reduction of mutation rates will retard the onset of generalized ageing or of a specific disease process.

There are, however, serious objections to the somatic mutation theory. The wasp *Habrobracon* reproduces parthenogenetically (i.e. without the need of sperm to fertilize the egg). It is possible to obtain individual wasps with either a diploid, or paired, set of chromosomes (as in most eukaryotes), or a haploid, single, set. Any gene mutation at an essential position in a haploid cell would result in loss of a vital process and impairment or death of the cell; in a diploid cell a serious mutation is often compensated for by the complementary gene and the cell can carry on its vital functions. Experiments have shown that haploid wasps live about as long as diploids, implying either that mutations are not a quantitatively important factor in ageing or that parthenogenetic species have compensated for the vulnerability of their haploids by developing an increased effectiveness of DNA repair.

Chromosomes can be separated into DNA and protein molecules, but this becomes more difficult in older cells. The isolated DNA of old animals, however, does not differ from that of the young. Although most of the DNA in a given cell at a given time is repressed (i.e. blocked from functioning), it is more repressed in old animals; it is not yet known whether this is a primary age change or a consequence of reduced cell metabolism arising from other causes.

PART 4

GENETICS TODAY AND TOMORROW

8

ANIMAL AND PLANT BREEDING

Animal Breeding

Introduction

Humanity has been modifying domesticated animals to better suit human needs for millennia. Selective breeding involves using knowledge from several branches of science, including genetics, statistics, reproductive physiology, computer science, and molecular genetics.

Breeding and Variation

British agriculturist Robert Bakewell was a very successful breeder of commercial livestock in the 18th century. His work was based on the traditional method of visual appraisal of the animals that he selected. Although he did not write about his methods, it is recorded that he travelled extensively around England and collected sheep and cattle that he considered useful. It is thought that he made wide outcrosses of diverse breeds, and then practised inbreeding with the intent of fixing desirable characteristics in the cross-bred animals. He was also the first to systematically let his animals for stud. For these reasons he is generally recognized as the first scientific breeder.

In animal breeding, a population is a group of interbreeding individuals (i.e. a breed or strain within a breed that is different in some aspects from other breeds or strains). Typically, certain animals within a breed are designated as pure-bred. The essential difference between pure-bred and non-pure-bred animals is that the genealogy of pure-bred animals has been carefully recorded, usually in a herd book, or stud book, kept by some sanctioning association. Pure-bred associations provide other services that are useful to their members to enhance their businesses.

Selective breeding utilizes the natural variations in traits that exist among members of any population. Breeding progress requires understanding the two sources of variation: genetics and environment. For some traits there is an interaction between genetics and the environment. Differences in the animals' environment, such as amount of feed, care, and even the weather, may have an impact on their growth, reproduction, and productivity. Such variations in performance because of the environment are not transmitted to the next generation. For most traits measured in domestic animals, the environment has a larger impact on variation than do genetic differences. For example, only about 30 per cent of the variation in milk production in dairy cattle can be attributed to genetic effects; the remainder of the variation is due to environmental effects. Thus, environmental factors must be considered and controlled in selecting breeding stock.

Genetic variation is necessary in order to make progress in breeding successive generations and is used for improving stock. Researchers partition total genetic variation into additive, dominance, and epistatic types of gene action, which are defined in the following paragraphs. Additive variation is easiest to use in breeding because it is common, and the effect of each allele at a locus just adds to the effect of other alleles at that same locus. Genetic gains made using additive genetic effects are permanent and cumulate from one generation to the next.

Although dominance variation is not more complex in theory, it is more difficult to control in practice because of how one allele masks the effect of another. For example, let a indicate a locus, with a_1 and a_2 representing two possible alleles at that location. Then a_1a_1, a_1a_2 (which is identical to a_2a_1), and a_2a_2 are the three possible genotypes. If a_1 dominates a_2, the genotypes a_1a_2 and a_1a_1 cannot be outwardly distinguished. Thus, the inability to observe differences between a_1a_2 and a_1a_1 presents a major difficulty in using dominance variance in selective breeding.

Additive and dominance variations are caused by genes at one locus. Epistatic variation is caused by the joint effects of genes at two or more loci. There has been little deliberate use of this type of genetic variation in breeding because of the complex nature of identifying and controlling the relevant genes.

Breeding

Breeding objectives

Breeding objectives can be discussed in terms of changing the genetic make-up of a population of animals, where a population is defined as a recognized breed. Choice of breeding goals and design of an effective breeding programme is usually not an easy task. Complicating the implementation of a breeding programme is the number of generations needed to reach the initial goals. Ultimately, breeding goals are dictated by market demand; however, it is not easy to predict what consumers will want several years in advance. Sometimes the marketplace demands a different product than was defined as desirable in the original breeding objective. When this happens, breeders have to adjust their programme, which results in less-efficient selection than if the new breeding goal had been used from the beginning. For example, consumers want beef that is leaner but still tender despite the lower fat content, and ranchers have changed their cattle-breeding programmes to meet this demand. The use of ultrasound is widespread in determining the fat and lean content of live animals, thus hastening the change of carcass quality to meet consumer demands. These trends have gradually changed over the last few decades; for example, Angus cattle are particularly noted for the quality of beef produced.

Additional complications arise from simultaneously trying to improve multiple traits and the difficulty of determining what part of the variation for each trait is under genetic control. In addition, some traits are genetically correlated, and this correlation may be positive or negative; i.e. the traits may be complementary or antagonistic. Breeding methods depend on heritability and genetic correlations for desirable traits.

Heritability and genetic correlations in breeding

Heritability is the proportion of the additive genetic variation to the total variation. Heritability is important because without genetic variation there can be no genetic change in the population. Alternatively, if heritability is high, genetic change can be quite rapid, and simple means of selection are all that is needed. Using an increasing scale from 0 to 1, a heritability of 0.75 means that 75 per cent of the total variance in a trait is controlled by additive gene action. With heritabilities as high as this, just the record of a single individual's traits can easily be used to create an effective breeding programme.

Some general statements can be made about heritability, keeping in mind that exceptions exist. Traits related to fertility have low heritabilities. Examples include the average number of times that a cow must be bred before she conceives and the average number of pigs in a litter. Traits related to production have intermediate heritabilities. Examples include the amount of milk a cow produces, the rates of weight gain in beef cattle and pigs, and the number of eggs laid by hens. So-called quality traits tend to have higher heritabilities. Examples include the amount of fat a pig has over its back and the amount of protein in a cow's milk. The magnitude of heritability is one of the primary considerations in designing breeding programmes.

Genetic correlation occurs when a single gene affects two traits. There may be many such genes that affect two or more traits. Genetic correlations can be positive or negative, which is indicated by assigning a number in the range from +1 to −1, with 0 indicating no genetic correlation. A correlation of +1 means that the traits always occur together, while a correlation of −1 means that having either trait always excludes having the other trait. Thus, the greater the displacement of the value from 0, the greater the correlation (positive or negative) between traits. The practical breeding consequence is that selection for one trait will pull along any positively correlated traits, even though there is no deliberate selection for them. For example, selecting for increased milk production also increases protein production. Another example is the selection for increased weight gain in broiler chickens, which also increases the fat content of the birds.

When traits have a negative genetic correlation, it is difficult to select for them simultaneously. For example, as milk production is increased in dairy cows through genetic selection, it is slightly more difficult for the high-producing cows to conceive. This negative correlation is partly due to the partitioning of the cows' nutrients between production and reproduction, with production being prioritized in early lactation. In the case of dairy cattle, milk production is in the order of 9,000 litres (20,000 pounds) per cow per year and is increasing. This is a large metabolic demand, so nutrient demand is large to meet this need. Thus, selecting for improved fertility may result in a reduction in milk production or its rate of gain.

Selection

Methods of selection

Types of selection are individual or mass selection, within and between family selection, sibling selection, and progeny testing, with many variations. Within family selection uses the best individual from each family for breeding. Between family selection uses the whole family for selection. Mass selection uses records of only the candidates for selection. Mass selection is most effective when heritability is high and the trait is expressed early in life, in which case all that is required is observation and selection based on phenotypes. When mass selection is not appropriate, other methods of selection, which make use of relatives or progeny, can be used singularly or in combination. Modern technologies allow use of all these types of selection at the same time, which results in greater accuracy.

Elements Needed to Make Genetic Progress

Genetic gain depends on balancing several factors: the accuracy of selection; the genetic variation in the population; the selection intensity (proportion selected for further breeding); and the generation interval (age of breeding). The genetic variation cannot be easily changed within a breed, though it can be changed by cross-breeding. The other factors can be changed. More complete pedigree records on candidates for breeding can increase the accuracy of selection, but waiting for candidates to reach full maturity in order to have better genetic data will increase the generation interval. Whether an increase in generation interval is justified by a more accurate selection process depends on individual circumstances. Selection intensity can also be increased, by narrowing the proportion of the population used in breeding, but it should be done without increasing the generation interval, which would have an unfavourable impact on genetic progress, all else being equal.

Evaluation of animals

Methods of ranking animals for breeding purposes have changed as statistical and genetic knowledge has increased. Along with increases in breeding knowledge, advancements in computing have enabled breeders to process routine breeding evaluations quickly and easily, as well as to develop research needed to rank large populations of animals. Evaluating and ranking candidates for selection depends on equating their performance record to a statistical model. A performance record can be calculated as the sum of genetic effects, known (categorized) environmental effects, and random environmental effects.

The first task in estimating the performance record is to statistically eliminate environmental effects, a process that involves setting up a system of equations to simultaneously solve for all of the genetic effects for the sires and dams. Information from relatives is included in the formula and increases the accuracy of evaluation of the candidates for selection. All relatives that are available can be incorporated in this type of evaluation. This model is called the animal model.

The animal model is used extensively in evaluating beef and dairy cattle, chickens, and pigs. To apply this model to evaluate large populations requires use of high-speed computers and extensive use of advanced mathematical techniques such as numerical analysis. In evaluating dairy cattle in the United States, a system of equations with more than 25 million variables is needed.

Accuracy of selection

In some cases the accuracy of selection for a trait can be measured using a calibrated tool or a scale. Thus, measurements of such traits can be replicated with high reliability. Some traits are difficult to measure on an objective scale, in which case a well-designed subjective scoring method can be effective. An excellent example is hip dysplasia, a degenerative disease of the hip joints that is common in many large dog breeds. Apparently, hip dysplasia is not associated with a single allele, making its incidence very difficult to control. However, an index has been developed by radiologists that allows young dogs to be assigned a score indicating their likelihood of developing the disease as they age. In 1997 American animal geneticist E.A. Leighton reported that, in fewer than five generations of selection in a breeding experiment using these scores, the incidence of canine hip dysplasia in German shepherd dogs measured at 12 to 16 months of age had decreased from the breed average of 55 per cent to 24 per cent among the experimental population; in Labrador retrievers the incidence dropped from 30 to 10 per cent.

Because close relatives share many genes, an examination of the relatives of a candidate for breeding can improve accuracy of selection. The more complete the genealogical record, or pedigree, the more effective the selection process. A pedigree is most useful when the heritability of a trait is relatively low, especially for traits that are expressed later in life or in only one sex.

Reproductive techniques can be used to increase the rate of genetic progress. In particular, for species that are mostly bred by artificial insemination, the best dams can be chosen and induced to superovulate, or release multiple eggs from their ovaries. These eggs are fertilized in the uterus and then flushed out in a non-surgical procedure that does not impair future conception of the donor female. Using this procedure, valuable females can produce more than one calf per year. Each embryo is implanted in a less-valuable host female to be carried through gestation. The sex of the embryos can be determined in utero at about 50 days of gestation. The normal gestation for Holstein–Friesian cattle is about 280 days, so this early determination of sex saves many days and allows the breeding programme to be adjusted. In particular, eggs from the donor cow could be collected again, or another superior cow could be bred to produce males. Thus, these reproductive technologies reduce the generation interval and increase selection intensity by getting more than one male calf from superior females. Both superovulation and sex determination are commonly used procedures. Superovulation is also used when breeders want to increase the number of female calves from a valuable cow.

Progeny testing

Progeny testing is used extensively in the beef and dairy cattle industry to aid in evaluating and selecting stock to be bred. Progeny testing is most useful when a high level of accuracy is needed for selecting a sire to be used extensively in artificial insemination. Progeny testing programmes consist of choosing the best sires and dams in the population based on an animal model evaluation, as described in the preceding section.

A description of progeny testing in dairy breeding provides a good example. The best 1 to 2 per cent of the cows from the population are chosen as bull mothers, and the best progeny-tested bulls are chosen to produce another generation of sires. The parents are mated to complement any individual deficiencies. The accuracy of evaluation of bull mothers is typically about 40 per cent, and for sires that produce young bulls the accuracy is more than 80 per cent. This is not as high as the industry wants for bulls to be used in artificial insemination. To reach greater accuracy, the next generation of sires is mated to enough cows in the population for each sire to produce 60 to 80 progeny. After the daughters of the young sires have a production record, the young sires are evaluated, and about the best 10 per cent are used extensively to produce commercial cows. Some of the progeny-tested sires will have thousands of daughters before a superior sire is found to replace them. About 70 per cent of dairy cattle are bred by artificial insemination, so these sires control the genetic destiny of dairy cattle. Consistently applying this selection procedure has been very successful.

The genetic gain has been consistent over the years. The actual first-lactation milk production varies more than the sire breeding value because differences in environmental conditions affect first-lactation production, but these environmental effects have been adjusted out of the breeding value calculations. There is no indication that the rate of gain in the sire breeding values is about to reduce. This level of achievement can only be realized if artificial insemination organizations and producers work together.

Breeding systems

Cross-breeding

Cross-breeding involves the mating of animals from two breeds. Normally, breeds are chosen that have complementary traits that will enhance the economic value of the offspring. An example is the cross-breeding of Yorkshire and Duroc breeds of pigs. Yorkshires have acceptable rates of gain in muscle mass and produce large litters, and Durocs are very muscular and have other acceptable traits, so these breeds are complementary. Another example is Angus and Charolais beef cattle. Angus produce high-quality beef and Charolais are especially large, so cross-breeding produces an animal with acceptable quality and size.

The other consideration in cross-breeding is heterosis, or hybrid vigour, which is displayed when the offspring performance exceeds the average performance of the parent breeds. This is a common phenomenon in which increased size, growth rate, and fertility are displayed by cross-bred offspring, especially when the breeds are more genetically dissimilar. Such increases generally do not increase in successive generations of cross-bred stock, so pure-bred lines must be retained for cross-breeding and for continual improvement in the parent breeds. In general, there is more heterosis for traits with low heritability. In particular, heterosis is thought to be associated with the collective action of many genes having small effects individually but large effects cumulatively. Because of hybrid vigour, a high proportion of commercial pork and beef comes from cross-bred animals.

Inbreeding

Mating animals that are related causes inbreeding. Inbreeding is often described as "narrowing the genetic base" because the mating of related animals results in offspring that have more genes in common. Inbreeding is used to concentrate desirable traits. Mild inbreeding has been used in some breeds of dogs and has been extensively used in laboratory mice and rats. For example, mice have been bred to be highly sensitive to compounds that might be detrimental or useful to humans. These mice are highly inbred so that researchers can obtain the same response with replicated treatments.

Inbreeding is generally detrimental in domestic animals. Increased inbreeding is accompanied by reduced fertility, slower growth rates, greater susceptibility to disease, and higher mortality rates. As a result, producers try to avoid mating related animals. This is not always possible, though, when long-continued selection for the same traits is practised within a small population, because parents of future generations are the best candidates from the last generation, and some inbreeding tends to accumulate. The rate of inbreeding can be reduced, but, if inbreeding depression becomes evident, some method of introducing more diverse genes will be needed. The most common method is some form of crossbreeding.

Plant Breeding

Introduction

Plant breeding is an ancient activity, dating to the very beginnings of agriculture. It is accomplished by selecting plants found to be economically or aesthetically desirable, first by controlling the mating of selected individuals, and then by selecting certain individuals among the progeny. Such processes, repeated over many generations, can change the hereditary make-up and value of a plant population far beyond the natural limits of previously existing populations.

Probably soon after the earliest domestications of cereal grains, humans began to recognize degrees of excellence among the plants in their fields and saved seed from the best for planting new crops. Such tentative selective methods were the forerunners of early plant-breeding procedures.

The results of early plant-breeding procedures were conspicuous. Most present-day varieties are so modified from their wild progenitors that they are unable to survive in nature. Indeed, in some cases, the cultivated forms are so strikingly different from existing wild relatives that it is difficult even to identify their ancestors. These remarkable transformations were accomplished by early plant breeders in a very short time from an evolutionary point of view, and the rate of change was probably greater than for any other evolutionary event.

Scientific plant breeding dates back hardly more than 50 years. The role of pollination and fertilization in the process of reproduction was not widely appreciated in the 19th century, and it was not until the early part of the 20th century that the laws of genetic inheritance were recognized and a beginning was made in applying them to the improvement of plants. One of the major facts that has emerged during the short history of scientific breeding is that an enormous wealth of genetic variability exists in the plants of the world and that only a start has been made in tapping its potential.

Goals

The plant breeder usually has in mind an ideal plant that combines a maximum number of desirable characteristics. These characteristics may include resistance to diseases and insects; tolerance to heat and frost; appropriate size, shape, and time to maturity; and many other general and specific traits that contribute to improved adaptation to the environment, ease in growing and handling, greater yield, and better quality. The breeder of fancy show plants must also consider aesthetic appeal. Thus, the breeder can rarely focus attention on any one characteristic but must take into account the manifold traits that make the plant more useful in fulfilling the purpose for which it is grown.

Increase of yield

One of the aims of virtually every breeding project is to increase yield. This can often be brought about by selecting obvious morphological variants. One example is the selection of dwarf, early maturing varieties of rice. These dwarf varieties are sturdy and give a greater yield of grain. Furthermore, their early maturity frees the land quickly, often allowing an additional planting of rice or another crop the same year.

Another way of increasing yield is to develop varieties resistant to diseases and insects. In many cases the development of resistant varieties has been the only practical method of pest control. Perhaps the most important feature of resistant varieties is the stabilizing effect they have on production and hence on steady food supplies. Varieties tolerant to drought, heat, or cold provide the same benefit.

Modifications of range and constitution

Another common goal of plant breeding is to extend the area of production of a crop species. A good example is the modification of grain sorghum since its introduction to the United States about 100 years ago. Of tropical origin, grain sorghum was originally confined to the southern Plains area and the Southwest, but earlier-maturing varieties were developed, which have become important as far north as North Dakota.

Development of crop varieties suitable for mechanized agriculture has become a major goal of plant breeding. Uniformity of plant characters is very important in mechanized agriculture because field operations are much easier when the individuals of a variety are similar in time of germination, growth rate, size of fruit, and so on. Uniformity in maturity is, of course, essential when crops such as tomatoes and peas are harvested mechanically.

The nutritional quality of plants can be greatly improved by breeding. For example, it is possible to breed varieties of corn (maize) that are much higher in the amino acid lysine than previously existing varieties. Breeding high-lysine maize varieties for those areas of the world where maize is the major source of this nutritionally essential amino acid has become a major goal in plant breeding.

In breeding ornamentals, attention is paid to such factors as longer blooming periods, improved keeping qualities of flowers, general thriftiness, and other features that contribute to usefulness and aesthetic appeal. Novelty itself is often a virtue in ornamentals, and the spectacular, even the bizarre, is often sought.

Evaluation of Plants

The appraisal of the value of plants so that the breeder can decide which individuals should be discarded and which allowed to produce the next generation is a much more difficult task with some traits than with others.

Qualitative characters

The easiest characters, or traits, to deal with are those involving discontinuous, or qualitative, differences that are governed by one or a few major genes. Many such inherited differences exist, and they frequently have profound effects on plant value and utilization. Examples are starchy versus sugary kernels (characteristic of field and sweet corn, respectively) and determinant versus indeterminant habit of growth in green beans (determinant varieties are adapted to mechanical harvesting). Such differences can be seen easily and evaluated quickly, and the expression of the traits remains the same regardless of the environment in which the plant grows. Traits of this type are termed highly heritable.

Quantitative characters

In other cases, however, plant traits grade gradually from one extreme to another in a continuous series, and classification into discrete classes is not possible. Such variability is termed quantitative. Many traits of economic importance are of this type; e.g. height, cold and drought tolerance, time to maturity, and, in particular, yield. These traits are governed by many genes, each having a small effect. Although the distinction between the two types of traits is not absolute, it is nevertheless convenient to designate qualitative characters as those involving discrete differences and quantitative characters as those involving a graded series.

Quantitative characters are much more difficult for the breeder to control, for three main reasons: (1) the sheer number of genes involved makes hereditary change slow and difficult to assess; (2) the variations of the traits involved are generally detectable only through measurement and exacting statistical analyses; and (3) most of the variations are due to the environment rather than to genetic endowment; for example, the heritability of certain traits is less than 5 per cent, meaning that 5 per cent of the observed variation is caused by genes and 95 per cent is caused by environmental influences.

It follows that carefully designed experiments are required to distinguish plants that are superior because they carry desirable genes from those that are superior because they happen to grow in a favourable site.

Methods of Plant Breeding

Mating systems

Plant mating systems devolve about the type of pollination, or transferral of pollen from flower to flower. A flower is self-pollinated (a "selfer") if pollen is transferred to it from any flower of the same plant and cross-pollinated (an "outcrosser" or "outbreeder") if the pollen comes from a flower on a different plant. About half of important cultivated plants are naturally cross-pollinated, and their reproductive systems include various devices that encourage cross-pollination; e.g. protandry (pollen shed before the ovules are mature, as in the carrot and walnut), dioecy (stamens and pistils borne on different plants, as in the date palm, asparagus, and hops), and genetically determined self-incompatibility (inability of pollen to grow on the stigma of the same plant, as in white clover, cabbage, and many other species).

Other plant species, including a high proportion of the most important cultivated plants, such as wheat, barley, rice, peas, beans, and tomatoes, are predominantly self-pollinating. There are relatively few reproductive mechanisms that promote self-pollination; the most positive is failure of the flowers to open (cleistogamy), as in certain violets. In barley, wheat, and lettuce the pollen is shed before or just as the flowers open; and in the tomato pollination follows opening of the flower, but the stamens form a cone around the stigma. In such species there is always a risk of unwanted cross-pollination.

In controlled breeding procedures it is imperative that pollen from the desired male parent, and no other pollen, reaches the stigma of the female parent. When stamens and pistils occur in the same flower, the anthers must be removed from flowers selected as females before pollen is shed. This is usually done with forceps or scissors. Protection must also be provided from "foreign" pollen. The most common method is to cover the flower with a plastic or paper bag. When the stigma of the female parent becomes receptive, pollen from the desired male parent is transferred to it, often by breaking an anther over the stigma, and the protective bag is replaced. The production of certain hybrids is, therefore, tedious and expensive because it often requires a series of delicate, exacting, and properly timed hand operations. When male and female parts occur in separate flowers, as in corn (maize), controlled breeding is easier.

A cross-pollinated plant, which has two parents, each of which is likely to differ in many genes, produces a diverse population of plants hybrid (heterozygous) for many traits. A self-pollinated plant, which has only one parent, produces a more uniform population of plants pure breeding (homozygous) for many traits. Thus, in contrast to outbreeders, self-breeders are likely to be highly homozygous and hence true breeding for a specified trait.

Breeding self-pollinated species

The breeding methods that have proved successful with self-pollinated species are: (1) mass selection; (2) pure-line selection; (3) hybridization, with the segregating generations handled by the pedigree method, the bulk method, or by the backcross method; and (4) development of hybrid varieties.

Mass selection

In mass selection, seeds are collected from (usually a few dozen to a few hundred) desirable-appearing individuals in a population, and the next generation is sown from the stock of mixed seed. This procedure, sometimes referred to as phenotypic selection, is based on how each individual looks. Mass selection has been used widely to improve old "land" varieties, i.e.

varieties that have been passed down from one generation of farmers to the next over long periods.

An alternative approach that has no doubt been practised for thousands of years is simply to eliminate undesirable types by destroying them in the field. The results are similar whether superior plants are saved or inferior plants are eliminated: seeds of the better plants become the planting stock for the next season.

A modern refinement of mass selection is to harvest the best plants separately and to grow and compare their progenies. The poorer progenies are destroyed and the seeds of the remainder are harvested. It should be noted that selection is based not solely on the appearance of the parent plants but also on the appearance and performance of their progeny. Progeny selection is usually more effective than phenotypic selection when dealing with quantitative characters of low heritability. However, progeny testing requires an extra generation; hence gain per cycle of selection must be double that of simple phenotypic selection to achieve the same rate of gain per unit time.

Mass selection, with or without progeny testing, is perhaps the simplest and least expensive of plant-breeding procedures. It finds wide use in the breeding of certain forage species, which are not important enough economically to justify more detailed attention.

Pure-line selection

Pure-line selection generally involves three more or less distinct steps: (1) numerous superior-appearing plants are selected from a genetically variable population; (2) progenies of the individual plant selections are grown and evaluated by simple observation, frequently over a period of several years; and (3) when selection can no longer be made on the basis of observation alone, extensive trials are undertaken, involving careful measurements to determine whether the remaining selections are superior in yielding ability and other aspects of performance. Any progeny superior to an existing variety is then released as a new "pure-line" variety.

Much of the success of this method during the early 1900s depended on the existence of genetically variable land varieties that were waiting to be exploited. They provided a rich source of superior pure-line varieties, some of which are still represented among commercial varieties. In recent years the pure-line method as outlined above has decreased in importance in the breeding of major cultivated species; however, the method is still widely used with the less important species that have not yet been heavily selected.

A variation of the pure-line selection method that dates back centuries is the selection of single-chance variants, mutations, or "sports" in the original variety. A very large number of varieties that differ from the original strain in characteristics such as colour, lack of thorns or barbs, dwarfness, and disease resistance have originated in this fashion.

Hybridization

During the 20th century planned hybridization between carefully selected parents became dominant in the breeding of self-pollinated species. The object of hybridization is to combine desirable genes found in two or more different varieties and to produce pure-breeding progeny superior in many respects to the parental types.

Gene alleles, however, are always in the company of other alleles and combine to form a genotype. The plant breeder's problem is largely one of efficiently managing the enormous numbers of genotypes that occur in the generations following hybridization. As an example of the power of hybridization in creating variability, a cross between hypothetical wheat varieties differing by only 21 genes is capable of producing more than 10,000 million different genotypes in the second generation. At spacings normally used by farmers, more than 50 million acres (about 200,000 km^2) would be required to grow a population large enough to permit every genotype to occur in its expected frequency. While the great majority of these second-generation genotypes are hybrid (heterozygous) for one or more traits, it is statistically possible that 2,097,152 different pure-breeding (homozygous) genotypes can occur, each potentially a new pure-line variety. These numbers illustrate the importance of efficient techniques in managing hybrid populations, for which purpose the pedigree procedure is most widely used.

Pedigree breeding starts with the crossing of two genotypes, each of which has one or more desirable characters lacked by the other. If the two original parents do not provide all the desired characters, a third parent can be included by crossing it to one of the hybrid progeny of the first generation (F_1). In the pedigree method superior types are selected in successive generations, and a record is maintained of parent–progeny relationships.

The F_2 generation (progeny of the crossing of two F_1 individuals) affords the first opportunity for selection in pedigree programmes. In this generation the emphasis is on the elimination of individuals carrying undesirable major genes. In the succeeding generations the hybrid condition gives way to pure breeding as a result of natural self-pollination, and families derived from different F_2 plants begin to display their unique character. Usually one or two superior plants are selected within each superior family in these generations. By the F_5 generation the pure-breeding condition (homozygosity) is extensive, and emphasis shifts almost entirely to selection between families. The pedigree record is useful in making these eliminations. At this stage each selected family is usually harvested in mass to obtain the larger amounts of seed needed to evaluate families for quantitative characters. This evaluation is usually carried out in plots grown under conditions that simulate commercial planting practice as closely as possible. When the number of families has been reduced to manageable proportions by visual selection, usually by the F_7 or F_8 generation, precise evaluation for performance and quality begins. The final evaluation of promising strains involves: (1) observation, usually in a number of years and locations, to detect weaknesses that may not have appeared previously; (2) precise yield testing; and (3) quality testing. Many plant breeders test for five years at five representative locations before releasing a new variety for commercial production.

The bulk-population method of breeding differs from the pedigree method primarily in the handling of generations following hybridization. The F_2 generation is sown at normal commercial planting rates in a large plot. At maturity the crop is harvested in mass, and the seeds are used to establish the next generation in a similar plot. No record of ancestry is kept. During the period of bulk propagation natural selection tends to eliminate plants having poor survival value. Two types of artificial selection are also often applied: (1) destruction of plants that carry undesirable major genes; and (2) mass techniques such as harvesting when only some of the seeds are mature to select for early maturing plants, or the use of screens to select for increased seed size. Single-plant selections are then made and evaluated in the same way as in the pedigree method of breeding. The chief advantage of the bulk population method is that it allows the breeder to handle very large numbers of individuals inexpensively.

Often an outstanding variety can be improved by transferring to it some specific desirable character that it lacks. This can be accomplished by first crossing a plant of the superior variety to a plant of the donor variety, which carries the trait in question, and then mating the progeny back to a plant having the genotype of the superior parent. This process is called backcrossing. After five or six backcrosses the progeny will be hybrid for the character being transferred but like the superior parent for all other genes. Selfing the last backcross generation, coupled with selection, will give some progeny pure-breeding for the genes being transferred. The advantages of the back-cross method are its rapidity, the small number of plants required, and the predictability of the outcome. A serious disadvantage is that the procedure diminishes the occurrence of chance combinations of genes, which sometimes leads to striking improvements in performance.

Hybrid varieties

The development of hybrid varieties differs from hybridization in that no attempt is made to produce a pure-breeding population; only the F_1 hybrid plants are sought. The F_1 hybrid of crosses between different genotypes is often much more vigorous than its parents. This hybrid vigour, or heterosis, can be manifested in many ways, including increased rate of growth, greater uniformity, earlier flowering, and increased yield, the last being of greatest importance in agriculture.

By far the greatest development of hybrid varieties has been in corn (maize), primarily because its male flowers (tassels) and female flowers (incipient ears) are separate and easy to handle, thus proving economical for the production of hybrid seed. The production of hand-produced F_1 hybrid seed of other plants, including ornamental flowers, has been economical only because greenhouse growers and home gardeners have been willing to pay high prices for hybrid seed.

However, a built-in cellular system of pollination control has made hybrid varieties possible in a wide range of plants, including many that are self-pollinating, such as sorghums. This system, called cytoplasmic male sterility, or cytosterility, prevents normal maturation or function of the male sex organs (stamens) and results in defective pollen or none at all. It obviates the need for removing the stamens either by hand or by machine. Cytosterility depends on the interaction between male sterile genes ($R + r$) and factors found in the cytoplasm of the female sex cell. The genes are derived from each parent in the normal Mendelian fashion, but the cytoplasm (and its factors) is provided by the egg only; therefore, the inheritance of cytosterility is determined by the female parent. All plants with fertile cytoplasm produce viable pollen, as do plants with sterile cytoplasm but at least one *R* gene; plants with sterile cytoplasm and two *r* genes are male sterile (produce defective pollen).

The production of F_1 hybrid seed between two strains is accomplished by interplanting a sterile version of one strain (A) in an isolated field with a fertile version of another strain (B). Since strain A produces no viable pollen, it will be pollinated by strain B, and all seeds produced on strain A plants must therefore be F_1 hybrids between the strains. The F_1 hybrid seeds are then planted to produce the commercial crop. Much of the breeder's work in this process is in developing the pure-breeding sterile and fertile strains to begin the hybrid seed production.

Breeding cross-pollinated species

The most important methods of breeding cross-pollinated species are: (1) mass selection; (2) development of hybrid varieties; and (3) development of synthetic varieties. Since cross-pollinated species are naturally hybrid (heterozygous) for many traits and lose vigour as they become pure-bred (homozygous), a goal of each of these breeding methods is to preserve or restore heterozygosity.

Mass selection

Mass selection in cross-pollinated species takes the same form as in self-pollinated species; i.e. a large number of superior-appearing plants are selected and harvested in bulk and the seed used to produce the next generation. Mass selection has proved to be very effective in improving qualitative characters, and, applied over many generations, it is also capable of improving quantitative characters, including yield, despite the low heritability of such characters. Mass selection has long been a major method of breeding cross-pollinated species, especially in the economically less important species.

Hybrid varieties

The outstanding example of the exploitation of hybrid vigour through the use of F_1 hybrid varieties has been with corn (maize). The production of a hybrid corn variety involves three steps: (1) the selection of superior plants; (2) selfing for several generations to produce a series of inbred lines, which although different from each other are each pure-breeding and highly uniform; and (3) crossing selected inbred lines. During the inbreeding process the vigour of the lines decreases drastically, usually to less than half that of field-pollinated varieties. Vigour is restored, however, when any two unrelated inbred lines are crossed, and in some cases the F_1 hybrids between inbred lines are much superior to open-pollinated varieties. An important consequence of the homozygosity of the inbred lines is that the hybrid between any two inbreds will always be the same. Once the inbreds that give the best hybrids have been identified, any desired amount of hybrid seed can be produced.

Pollination in corn (maize) is by wind, which blows pollen from the tassels to the styles (silks) that protrude from the tops of the ears. Thus controlled cross-pollination on a field scale can be accomplished economically by interplanting two or three rows of the seed parent inbred with one row of the pollinator, inbred and de-tasselling the former before it sheds pollen. In practice, most hybrid corn is produced from "double crosses", in which four inbred lines are first crossed in pairs (A × B and C × D) and then the two F_1 hybrids are crossed again, (A × B) × (C × D). The double-cross procedure has the advantage that the commercial F_1 seed is produced on the highly productive single cross A × B rather than on a poor-yielding inbred, thus reducing seed costs. In recent years cytoplasmic male sterility, described earlier, has been used to eliminate de-tasselling of the seed parent, thus providing further economies in producing hybrid seed.

Much of the hybrid vigour exhibited by F_1 hybrid varieties is lost in the next generation. Consequently, seed from hybrid varieties is not used for planting stock; the farmer purchases new seed each year from seed companies.

Perhaps no other development in the biological sciences has had greater impact on increasing the quantity of food supplies available to the world's population than has the development of hybrid corn (maize). Hybrid varieties in other crops, made possible through the use of male sterility, have also been dramatically successful, and it seems likely that use of hybrid varieties will continue to expand in the future.

Synthetic varieties

A synthetic variety is developed by intercrossing a number of genotypes of known superior combining ability – i.e. genotypes that are known to give superior hybrid performance when crossed in all combinations. (By contrast, a variety developed by mass selection is made up of genotypes bulked together without having undergone preliminary testing to determine their performance in hybrid combination.) Synthetic varieties are known for their hybrid vigour and for their ability to produce usable seed for succeeding seasons. Because of these advantages, synthetic varieties have become increasingly favoured in the growing of many species, such as the forage crops, in which expense prohibits the development or use of hybrid varieties.

Disease Resistance in Plants

Host Resistance and Selection

Disease-resistant varieties of plants offer an effective, safe, and relatively inexpensive method of control for many crop diseases. Most available commercial varieties of crop plants bear resistance to at least one, and often several, pathogens. Resistant or immune varieties are critically important for low-value crops in which other controls are unavailable or impractical because they are too expensive. Much has been accomplished in developing disease-resistant varieties of field crops, vegetables, fruits, turf grasses, and ornamentals. Although great flexibility and potential for genetic change exist in most economically important plants, pathogens are also flexible. Sometimes, a new plant variety is developed that is highly susceptible to a previously unimportant pathogen.

Variable resistance

Resistance to disease varies among plants; it may be either total (a plant is immune to a specific pathogen) or partial (a plant is tolerant to a pathogen, suffering minimal injury). The two broad categories of resistance to plant diseases are vertical (specific) and horizontal (non-specific). A plant variety that exhibits a high degree of resistance to a single race, or strain, of a pathogen is said to be vertically resistant; this ability usually is controlled by one or a few plant genes. Horizontal resistance, on the other hand, protects plant varieties against several strains of a pathogen, although the protection is not as complete. Horizontal resistance is more common and involves many genes.

Obtaining disease-resistant plants

Several means of obtaining disease-resistant plants are commonly employed alone or in combination. These include introduction from an outside source, selection, and induced variation. All three may be used at different stages in a continuous process; for example, varieties free from injurious insects or plant diseases may be introduced for comparison with local varieties. The more promising lines or strains are then selected for further propagation, and they are further improved by promoting as much variation as possible through hybridization or special treatment. Finally, selection of the plants showing greatest promise takes place. Developing disease-resistant plants is a continuing process.

Special treatments for inducing gene changes include the application of mutation-inducing chemicals and irradiation with ultraviolet light and X-rays. These treatments commonly induce deleterious genetic changes, but beneficial changes may also occur occasionally.

Methods used in breeding plants for disease resistance are similar to those used in breeding for other characters, except that two organisms are involved – the host plant and the pathogen. Thus, it is necessary to know as much as possible about the nature of inheritance of the resistant characters in the host plant and the existence of physiological races or strains of the pathogen.

9

BIOTECHNOLOGY

Introduction

Biotechnology is the use of biology to solve problems and make useful products. The most prominent area of biotechnology is the production of therapeutic proteins and other drugs through genetic engineering.

People have been harnessing biological processes to improve their quality of life for some 10,000 years, beginning with the first agricultural communities. Approximately 6,000 years ago, humans began to tap the biological processes of microorganisms in order to make bread, alcoholic beverages, and cheese, and to preserve dairy products. But such processes are not what is meant today by *biotechnology*, a term first widely applied to the molecular and cellular technologies that began to emerge in the 1960s and 1970s. A fledgling "biotech" industry began to coalesce in the mid- to late 1970s, led by Genentech, a pharmaceutical company established in 1976 by American chemist and venture capitalist Robert A. Swanson and American biologist Herbert W. Boyer to commercialize the recombinant DNA technology pioneered by Boyer and American biochemist Stanley N. Cohen. Early companies such as Genentech, Amgen, Biogen, Cetus, and Genex began by manufacturing genetically engineered substances primarily for medical and environmental uses.

For more than a decade, the biotechnology industry was dominated by recombinant DNA technology. This technique consists of splicing the gene for a useful protein (often a human protein) into production cells – such as yeast, bacteria, or mammalian cells in culture – which then begin to produce the protein in volume. In the process of splicing a gene into a production cell, a new organism is created.

At first, biotechnology investors and researchers were uncertain about whether the courts would permit them to acquire patents on organisms; after all, patents were not allowed on new organisms that happened to be discovered and identified in nature. But, in 1980, the US Supreme Court, in the case of *Diamond* v. *Chakrabarty*, resolved the matter by ruling that "a live human-made microorganism is patentable matter". This decision spawned a wave of new biotechnology firms and the infant industry's first investment boom. In 1982 recombinant insulin became the first product made through genetic engineering to secure approval from the US Food and Drug Administration (FDA). Since then, dozens of genetically engineered protein medications have been commercialized around the world, including recombinant versions of growth hormone, clotting factors, proteins for stimulating the production of red and white blood cells, interferons, and clot-dissolving agents.

In the early years, the main achievement of biotechnology was the ability to produce naturally occurring therapeutic molecules in larger quantities than could be derived from conventional sources, such as plasma, animal organs, and human cadavers. Recombinant proteins are also less likely to be contaminated with pathogens or to provoke allergic reactions. Today, biotechnology researchers seek to discover the root molecular causes of disease and to intervene precisely at that level. Sometimes this means producing therapeutic proteins that augment the body's own supplies or that make up for genetic deficiencies, as in the first generation of biotech medications. (Gene therapy is a related approach – see below.) But the biotechnology industry has also expanded its research into the development of traditional pharmaceuticals and monoclonal antibodies that stop the progress of a disease. Such steps are uncovered through painstaking study of genes (genomics), the proteins that they encode (proteomics), and the larger biological pathways in which they act.

In addition to the tools mentioned above, biotechnology also involves merging biological information with computer technology (bioinformatics), exploring the use of microscopic equipment that can enter the human body (nanotechnology), and possibly applying techniques of stem cell research and cloning to replace dead or defective cells and tissues (regenerative medicine). Companies and academic laboratories integrate these disparate technologies in an effort to analyse downwards into molecules and also to synthesize upwards from molecular biology toward chemical pathways, tissues, and organs.

In addition to being used in health care, biotechnology has proved helpful in refining industrial processes through the discovery and production of biological enzymes that spark chemical reactions (catalysts); for environmental clean-up, with enzymes that digest contaminants into harmless chemicals and then die after consuming the available "food supply"; and in agricultural production through genetic engineering.

Agricultural applications of biotechnology have proved the most controversial. Some activists and consumer groups have called for bans on genetically modified organisms (GMOs) or for labelling laws to inform consumers of the growing presence of GMOs in the food supply. The introduction of GMOs into agriculture began in 1993, when the FDA approved bovine somatotropin (BST), a growth hormone that boosts milk production in dairy cows. The next year, the FDA approved the first genetically modified whole food, a tomato engineered for a longer shelf life. Since then, regulatory approval in the United States, Europe, and elsewhere has been won by dozens of agricultural GMOs, including crops that produce their own pesticides and crops that survive the application of specific herbicides used to kill weeds. Studies by the United Nations, the US National Academy of Sciences, the European Union, the American Medical Association, US regulatory agencies, and other organizations have found GMO foods to be safe, but sceptics contend that it is still too early to judge the long-term health and ecological effects of such crops. In the late 20th and early 21st centuries, the land area planted in genetically modified crops increased dramatically; for example, between 1997 and 2002, the area allocated to genetically modified crops increased 30-fold. (See Chapter 10 for a more detailed discussion of GMOs.)

Overall, the revenues of US and European biotechnology industries roughly doubled over the five-year period from 1996 to 2000. Rapid growth continued into the 21st century, fuelled by the introduction of new products, particularly in health care.

Recombinant DNA Technology

Recombinant DNA technology is the joining together of DNA molecules from two different species that are inserted into a host organism to produce new genetic combinations that are of value to science, medicine, agriculture, and industry.

Since the focus of all genetics is the gene, the fundamental goal of laboratory geneticists is to isolate, characterize, and manipulate genes. Although it is relatively easy to isolate a sample of DNA from a collection of cells, finding a specific gene within this DNA sample can be compared to finding a needle in a haystack. Consider the fact that each human cell contains approximately 2 metres (6 feet) of DNA. Therefore, a small tissue sample will contain many kilometres of DNA. However, recombinant DNA technology has made it possible to isolate one gene or any other segment of DNA, enabling researchers to determine its nucleotide sequence, study its transcripts, mutate it in highly specific ways, and re-insert the modified sequence into a living organism.

DNA Cloning

In biology a clone is a group of individual cells or organisms descended from one progenitor. This means that the members of a clone are genetically identical, because cell replication produces identical daughter cells each time. The use of the word "clone" has been extended to recombinant DNA technology, which has provided scientists with the ability to produce many copies of a single fragment of DNA, such as a gene, creating identical copies that constitute a DNA clone. In practice, the procedure is carried out by inserting a DNA fragment into a small DNA molecule and then allowing this molecule to replicate inside a simple living cell such as a bacterium. The small replicating molecule is called a DNA vector (carrier).

Vectors

The most commonly used vectors are plasmids (circular DNA molecules that originated from bacteria), viruses, and yeast cells. Plasmids are not a part of the main cellular genome, but they can carry genes that provide the host cell with useful properties, such as drug resistance, mating ability, and toxin production. They are small enough to be conveniently manipulated experimentally, and, furthermore, they will carry extra DNA that is spliced into them

Several bacterial viruses have also been used as vectors. The most commonly used is the lambda phage. The central part of the lambda genome is not essential for the virus to replicate in *Escherichia coli*, so this can be excised using an appropriate restriction enzyme, and inserts from donor DNA can be spliced into the gap. In fact, when the phage repackages DNA into its protein capsule, it includes only DNA fragments the same length as the normal phage genome.

Vectors are chosen depending on the total amount of DNA that must be included in a library. Cosmids are engineered vectors that are hybrids of plasmid and phage lambda; however, they can carry larger inserts than either pUC plasmids (plasmids engineered to produce a very high number of DNA copies but that can accommodate only small inserts) or lambda phage alone. Bacterial artificial chromosomes (BACs) are vectors based on F-factor (fertility factor) plasmids of *E. coli* and can carry much larger amounts of DNA. Yeast artificial chromosomes (YACs) are vectors based on autonomously replicating plasmids of *Saccharomyces cerevisiae* (baker's yeast). In yeast, which is a eukaryotic organism, a YAC behaves like a yeast chromosome and segregates properly into daughter cells. These vectors can carry the largest inserts of all and are used extensively in cloning large genomes such as the human genome.

Creating the clone

The steps in cloning are as follows (Figure 9.1). DNA is extracted from the organism under study and is cut into small fragments of a size suitable for cloning. Most often this is achieved by cleaving the DNA with a restriction enzyme. Restriction enzymes are extracted from several different species and strains of bacteria, in which they act as defence mechanisms against viruses. They can be thought of as "molecular scissors", cutting the DNA at specific target sequences. The most useful restriction enzymes make staggered cuts; that is, they leave a single-stranded overhang at the site of cleavage. These overhangs are very useful in cloning because the unpaired nucleotides will pair with other overhangs made using the same restriction enzyme. So, if the donor DNA and the vector DNA are both cut with the same enzyme, there is a strong possibility that the donor fragments and the cut vector will splice together because of the complementary overhangs. The resulting molecule is called recombinant DNA. It is recombinant in the sense that it is composed of DNA from two different sources. Thus, it is a type of DNA that would not occur naturally and is an artefact created by DNA technology.

The next step in the cloning process is to cut the vector with the same restriction enzyme used to cut the donor DNA. Vectors have target sites for many different restriction enzymes, but the most convenient ones are those that occur only once in the vector molecule. This is because the restriction enzyme then merely opens up the vector ring, creating a space for the insertion of the donor DNA segment. Cut vector DNA and donor DNA are mixed in a test tube, and the complementary ends of both types of DNA unite randomly. Of course, several types of unions are possible: donor fragment to donor fragment; vector fragment to vector fragment; and, most important, vector fragment to donor fragment, which can be selected for. Recombinant DNA associations form spontaneously in the above manner, but these associations are not stable because, although the ends are paired, the sugar–phosphate backbone of the DNA has not been sealed. This is

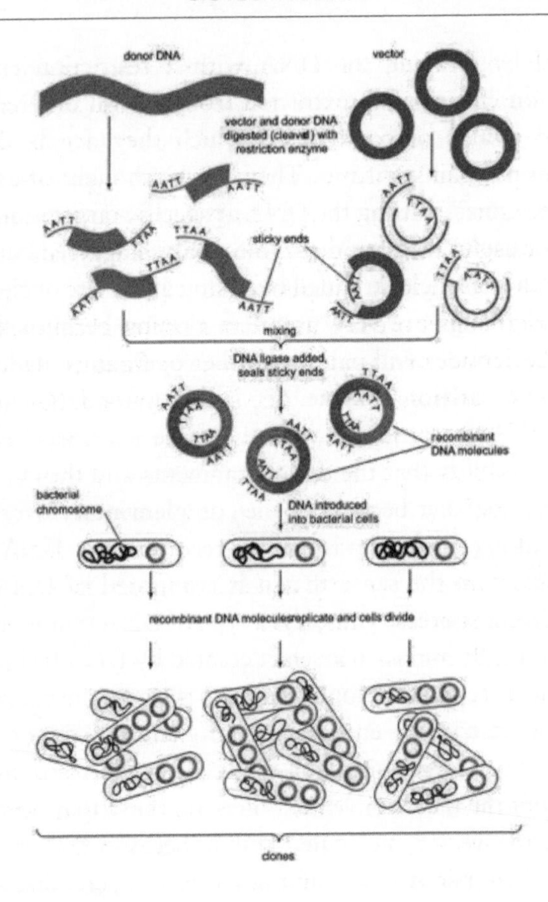

Figure 9.1 Steps involved in the engineering of a recombinant DNA molecule.

accomplished by the application of an enzyme called DNA ligase, which seals the two segments, forming a continuous and stable double helix.

The mixture should now contain a population of vectors each containing a different donor insert. This solution is mixed with live bacterial cells that have been specially treated to make their cells more permeable to DNA. Recombinant molecules enter living cells in a process called transformation. Usually, only a single recombinant molecule will enter any individual bacterial cell. Once inside, the recombinant DNA molecule replicates like any other plasmid DNA molecule, and many copies are subsequently produced. Furthermore, when the bacterial cell divides, all of the daughter cells receive the recombinant plasmid, which again replicates in each daughter cell.

The original mixture of transformed bacterial cells is spread out on the surface of a growth medium in a flat dish (Petri dish) so that the cells are separated from one another. These individual cells are invisible to the naked eye, but as each cell undergoes successive rounds of cell division, visible colonies form. Each colony is a cell clone, but it is also a DNA clone because the recombinant vector has now been amplified by replication during every round of cell division. Thus, the Petri dish may support many hundreds of distinct colonies, representing a large number of clones of different DNA fragments. This collection of clones is called a DNA library. By considering the size of the donor genome and the average size of the inserts in the recombinant DNA molecule, a researcher can calculate the number of clones needed to encompass the entire donor genome, or, in other words, the number of clones needed to constitute a genomic library.

Another type of library is a cDNA library. Creation of a cDNA library begins with messenger RNA (mRNA) instead of DNA. Messenger RNA carries encoded information from DNA to ribosomes for translation into protein. To create a cDNA library, these mRNA molecules are treated with the enzyme reverse transcriptase, which is used to make a DNA copy of an mRNA. The resulting DNA molecules are called complementary DNA (cDNA). A cDNA library represents a sampling of the transcribed genes, whereas a genomic library includes untranscribed regions.

Both genomic and cDNA libraries are made without regard to obtaining functional cloned donor fragments. Genomic clones do not necessarily contain full-length copies of genes. Furthermore, genomic DNA from eukaryotes contains introns, which are regions of DNA that are not translated into protein and cannot be processed by bacterial cells. This means that even full-sized genes are not translated in their entirety. In addition, eukaryotic regulatory signals are different from those used by prokaryotes. However, it is possible to produce expression libraries by slicing cDNA inserts immediately adjacent to a bacterial promoter region on the vector. In these expression libraries, eukaryotic proteins are made in bacterial cells, which allows several important technological applications that are discussed below in the section on DNA sequencing.

Isolating the clone

In general, cloning is undertaken in order to obtain the clone of one particular gene or DNA sequence of interest. The next step after cloning, therefore, is to find and isolate that clone among other members of the library. If the library encompasses the whole genome of an organism, then somewhere within that library will be the desired clone. There are several ways of finding it, depending on the specific gene concerned. Most commonly, a cloned DNA segment that shows homology to the sought gene is used as a probe. For example, if a mouse gene has already been cloned, then that clone can be used to find the equivalent human clone from a human genomic library. Bacterial colonies constituting a library are grown in a collection of Petri dishes. Then a porous membrane is laid over the surface of each plate, and cells adhere to the membrane. The cells are ruptured, and DNA is separated into single strands – all on the membrane. The probe is also separated into single strands and labelled, often with radioactive phosphorus. A solution of the radioactive probe is then used to bathe the membrane. The single-stranded probe DNA will adhere only to the DNA of the clone that contains the equivalent gene. The membrane is dried and placed against a sheet of radiation-sensitive film, and somewhere on the film a black spot will appear, announcing the presence and location of the desired clone. The clone can then be retrieved from the original Petri dishes.

DNA Sequencing

Uses

Knowledge of the sequence of a DNA segment has many uses, and some examples follow.

Finding genes

If a region of DNA has been sequenced, it can be screened for characteristic features of genes. For example, open reading frames – long sequences that begin with a start codon (three adjacent nucleotides; the sequence of a codon dictates amino acid production) and are uninterrupted by stop codons (except for one at their termination) – suggest a protein-coding region (see Chapter 5, Gene mutations). Also, human genes are generally adjacent to so-called CpG islands – clusters of cytosine and guanine, two of the nucleotides that make up DNA. If a gene with a known phenotype (such as a disease gene in humans) is known to be in the chromosomal region sequenced, then unassigned genes in the region will become candidates for that function.

Plotting evolutionary relationships

Homologous DNA sequences of different organisms can be compared in order to plot evolutionary relationships both within and between species. (See Figure 4.4 and detailed discussion in Chapter 5, DNA and Protein as Informational Macromolecules.)

Determining the function of a gene
A gene sequence can be screened for functional regions. In order to determine the function of a gene, various domains can be identified that are common to proteins of similar function. For example, certain amino acid sequences within a gene are always found in proteins that span a cell membrane; such amino acid stretches are called transmembrane domains. If a transmembrane domain is found in a gene of unknown function, it suggests that the encoded protein is located in the cellular membrane. Other domains characterize DNA-binding proteins. Several public databases of DNA sequences are available for analysis by any interested researcher.

Methods

The two basic gene sequencing approaches are the Maxam–Gilbert method, discovered by and named for American molecular biologists Allan M. Maxam and Walter Gilbert, and the Sanger method, discovered by British biochemist Frederick Sanger. In the most commonly used method, the Sanger method, DNA chains are synthesized on a template strand, but chain growth is stopped by the addition of a chemically modified form of the A, C, G, or T nucleotide. A population of nested, truncated DNA molecules results that represents each of the sites of that particular nucleotide in the template DNA. These molecules are separated by electrophoresis, and the inferred nucleotide sequence is deduced using computer analysis.

In Vitro Mutagenesis

Another use of cloned DNA is in vitro mutagenesis in which a mutation is produced in a segment of cloned DNA. The mutated DNA is then inserted into a cell or organism, and the effects of the mutation are studied. Mutations help geneticists to investigate the components of any biological process. However, traditional mutational analysis relied on the occurrence of random spontaneous mutations – a hit-or-miss method in which it was impossible to predict the precise type or position of the mutations obtained. In vitro mutagenesis, however, allows specific mutations to be tailored for type and for position within the gene. A cloned gene is treated in the test tube (in vitro) to obtain the specific mutation desired, and then this fragment is reintroduced into the living cell, where it replaces the resident gene.

One method of in vitro mutagenesis is oligonucleotide-directed mutagenesis. A specific point in a sequenced gene is pinpointed for mutation. An oligonucleotide, a short stretch of synthetic DNA of the desired sequence, is made chemically. For example, the oligonucleotide might have adenine (A) in one specific location instead of guanine (G). This oligonucleotide is hybridized to the complementary strand of the cloned gene; it will hybridize despite the mismatch of one base pair. Various enzymes are added to allow the oligonucleotide to prime the synthesis of a complete strand within the vector. When the vector is introduced into a bacterial cell and replicates, the mutated strand will act as a template for a complementary strand that will also be mutant, and thus a fully mutant molecule is obtained. This fully mutant cloned molecule is then reintroduced into the donor organism, and the mutant DNA replaces the resident gene.

Another version of in vitro mutagenesis is gene disruption, or gene knockout. In this technique the resident functional gene is replaced by a completely non-functional copy. The advantage of this technique over random mutagenesis is that specific genes can be knocked out at will, leaving all other genes untouched by the mutagenic procedure.

Genetically Modified Organisms

The ability to obtain specific DNA clones using recombinant DNA technology has made it possible to add the DNA of one organism to the genome of another. The added gene is called a transgene. The transgene inserts itself into a chromosome and is passed to the progeny as a new component of the genome. The resulting organism carrying the transgene is called a transgenic organism or genetically modified organism (GMO). In this way, a "designer organism" is made that contains some specific change required for an experiment in basic genetics or for improvement of some commercial strain. Several transgenic plants have been produced. Genes for toxins that kill insects have been introduced in several species, including corn and cotton. Bacterial genes that confer resistance to herbicides also have been introduced into crop plants. Other plant transgenes aim at improving the nutritional value of the plant.

Gene Therapy

Gene therapy is the introduction of a normal gene into an individual's genome in order to repair a mutation that causes a genetic disease. When a normal gene is inserted into a mutant nucleus, it most likely will integrate into a chromosomal site different from the defective allele; although this may repair the mutation, a new mutation may result if the normal gene integrates into another functional gene. If the normal gene replaces the mutant allele, there is a chance that the transformed cells will proliferate and produce enough normal gene product for the entire body to be restored to the non-diseased phenotype. So far, human gene therapy has been attempted only on somatic (body) cells for diseases such as cancer and severe combined immunodeficiency syndrome (SCIDS). Somatic cells cured by gene therapy may reverse the symptoms of disease in the treated individual, but the modification is not passed on to the next generation. Germinal gene therapy aims to place corrected cells inside the germ line (e.g. cells of the ovary or testis). If this is achieved, these cells will undergo meiosis and provide a normal gametic contribution to the next generation. Germinal gene therapy has been achieved experimentally in animals but not in humans.

Reverse Genetics

Recombinant DNA technology has made possible a type of genetics called reverse genetics. Traditionally, genetic research starts with a mutant phenotype, and, by Mendelian crossing analysis, a researcher is able to attribute the phenotype to a specific gene. Reverse genetics travels in precisely the opposite direction. Researchers begin with a gene of unknown function and use molecular analysis to determine its phenotype. One important tool in reverse genetics is gene knockout. By mutating the cloned gene of unknown function and using it to replace the resident copy or copies, the resultant mutant phenotype will show which biological function this gene normally controls.

Diagnostics

Recombinant DNA technology has led to powerful diagnostic procedures useful in both medicine and forensics. In medicine these diagnostic procedures are used in counselling prospective parents as to the likelihood of having a child with a particular disease, and they are also used in the prenatal prediction of genetic disease in the fetus. Researchers look for specific DNA fragments that are located in close proximity to the gene that causes the disease of concern. These fragments, called restriction fragment length polymorphisms, often serve as effective "genetic markers".

DNA Fingerprinting

This technique was developed in 1984 by the British geneticist Alec Jeffreys, after he noticed the existence of certain sequences of DNA (called minisatellites) that do not contribute to the function of a gene but are repeated within the gene and in other genes of a DNA sample. Jeffreys also determined that each organism has a unique pattern of these minisatellites, the only exception being multiple individuals from a single zygote (e.g. identical twins).

The procedure for creating a DNA fingerprint consists of first obtaining a sample of cells containing DNA (e.g. from skin, blood, or hair), extracting the DNA, and purifying it. The DNA is then cut at specific points along the strand by restriction enzymes. This produces fragments of varying lengths that are sorted by placing them on a gel and then subjecting the gel to electrophoresis: the shorter the fragment, the more quickly it will move toward the positive pole (anode). The sorted, double-stranded DNA fragments are then subjected to a blotting technique in which they are split into single strands and transferred to a nylon sheet. The fragments undergo autoradiography in which they are exposed to DNA probes – pieces of synthetic DNA that have been made radioactive and that bind to the minisatellites. A piece of X-ray film is then exposed to the fragments, and a dark mark is produced at any point where a radioactive probe has become attached. The resultant pattern of these marks can then be analysed.

An early use of DNA fingerprinting was in legal disputes, notably to help solve crimes and to determine paternity. The technique was challenged, however, over concerns about sample contamination, faulty preparation procedures, and erroneous interpretation of the results. Efforts have been made to improve reliability.

If only a small amount of DNA is available for fingerprinting, the PCR technique may be used to create thousands of copies of a DNA segment. PCR is an automated procedure in which certain oligonucleotide primers are used to repeatedly duplicate specific segments of DNA. Once an adequate amount of DNA has been produced, the exact sequence of nucleotide pairs in a segment of DNA can be determined using one of several biomolecular sequencing methods.

New automated equipment has greatly increased the speed of DNA sequencing and made available many new practical applications, including pinpointing segments of genes that cause genetic diseases, mapping the human genome, engineering drought-resistant plants, and producing biological drugs from genetically altered bacteria.

Protein Manufacture

Recombinant DNA procedures have been used to convert bacteria into "factories" for the synthesis of foreign proteins. This technique is useful not only for preparing large amounts of protein for basic research but also for producing valuable proteins for medical use. For example, the genes for human proteins such as growth hormone, insulin, and blood-clotting factors can be commercially manufactured. Another approach to producing proteins via recombinant DNA technology is to introduce the desired gene into the genome of an animal, engineered in such a way that the protein is secreted in the animal's milk, facilitating harvesting.

DNA Vaccines

In the late 20th century, advances in laboratory techniques allowed approaches to vaccine development to be refined. Medical researchers could identify the genes of a pathogen that encode antigen – the protein or proteins that stimulate the immune response to that organism. This has allowed the antigens to be mass-produced and used in vaccines. It has also made it possible to alter pathogens genetically and produce weakened strains of viruses. In this way, harmful proteins from pathogens can be deleted or modified, thus providing a safer and more effective method by which to manufacture attenuated vaccines.

Recombinant DNA technology has also proved useful in developing vaccines to viruses that cannot be grown successfully or that are inherently dangerous. Genetic material that codes for a desired antigen is inserted into the attenuated form of a large virus, such as the vaccinia virus, which carries the foreign genes "piggyback". The altered virus is injected into an individual to stimulate antibody production to the foreign proteins and thus confer immunity. This approach potentially enables the vaccinia virus to function as a live vaccine against several diseases, once it has received genes derived from the relevant disease-causing microorganisms. A similar procedure can be followed using a modified bacterium, such as *Salmonella typhimurium*, as the carrier of a foreign gene.

Another approach, called naked DNA therapy, involves injecting DNA that encodes a foreign protein into muscle cells. These cells produce the foreign antigen, which stimulates an immune response.

Genomics

Genomics is the study of the structure, function, and inheritance of the genome of an organism. A major part of genomics is determining the sequence of molecules that make up the DNA content of an organism. Every organism contains a basic set of chromosomes, unique in number and size for every species, that includes the complete set of genes plus any DNA between them. Although the term "genome" was not brought into use until 1920, the existence of chromosomes had been known since the late 19th century, when they were first observed as stained bodies visible under the microscope.

The initial discovery of chromosomes was followed in the 20th century by the mapping of genes on chromosomes based on the frequency of exchange of parts of chromosomes by a process called chromosomal crossing-over, an event that occurs as a part of the normal process of recombination and the production of sex cells (gametes) during meiosis. The genes that could be mapped by chromosomal crossing-over were mainly those for which mutant phenotypes (visible manifestations of an organism's genetic composition) had been observed, only a small proportion of the total genes in the genome. The discipline of genomics arose when the technology became available to deduce the complete nucleotide sequence of genomes, sequences generally in the range of billions of nucleotide pairs.

Sequencing and Bioinformatic Analysis of Genomes

Genomic sequences are usually determined using automatic sequencing machines. In a typical experiment to determine a genomic sequence, genomic DNA is first extracted from a sample of cells of an organism and then broken into many random fragments. These fragments are cloned in a DNA vector that is capable of carrying large DNA inserts (see Vectors, above).

Because the total amount of DNA that is required for sequencing and additional experimental analysis is several times the total amount of DNA in an organism's genome, each of the cloned fragments is amplified individually by replication inside a living bacterial cell, which reproduces rapidly and in great quantity to generate many bacterial clones. The cloned DNA is then extracted from the bacterial clones and is fed into the sequencing machine. The resulting sequence data are stored in a computer. When a large enough number of sequences from many different clones is obtained, the computer ties them together using sequence overlaps. The result is the genomic sequence, which is then deposited in a publicly accessible database. (For more information about DNA cloning and sequencing, see the section above on Recombinant DNA Technology.)

A complete genomic sequence in itself is of limited use; the data must be processed to find the genes and, if possible, their associated regulatory sequences. The need for these detailed analyses has given rise to the field of bioinformatics, in which computer programs scan DNA sequences looking for genes, using algorithms based on the known features of genes, such as unique triplet sequences of nucleotides known as start and stop codons that span a gene-sized segment of DNA, or sequences of DNA that are known to be important in regulating adjacent genes.

Once candidate genes are identified, they must be annotated to ascribe potential functions. Such annotation is generally based on known functions of similar gene sequences in other organisms, a type of analysis made possible by evolutionary conservation of gene sequence and function across organisms as a result of their common ancestry. However, after annotation there are still some genes whose functions cannot be deduced and must be revealed by further research.

Genomics Applications

Functional genomics

Analysis of genes at the functional level is one of the main uses of genomics, an area known generally as functional genomics. Determining the function of individual genes can be done in several ways. Classical, or forward, genetic methodology starts with a randomly obtained mutant of interesting phenotype and uses this to find the normal gene sequence and its function. Reverse genetics starts with the normal gene sequence (as obtained by genomics), induces a targeted mutation into the gene, then, by observing how the mutation changes phenotype, deduces the normal function of the gene.

The two approaches, forward and reverse, are complementary. Often a gene identified by forward genetics has been mapped to one specific chromosomal region, and the full genomic sequence reveals a gene in this position with an already annotated function.

Gene identification by microarray genomic analysis

Genomics has greatly simplified the process of finding the complete subset of genes that is relevant to some specific temporal or developmental event of an organism. For example, microarray technology allows a sample of the DNA of a clone of each gene in a whole genome to be laid out in order on the surface of a special chip, which is basically a small thin piece of glass that is treated in such a way that DNA molecules stick firmly to the surface. For any specific developmental stage of interest (e.g. the growth of root hairs in a plant or the production of a limb bud in an animal), the total RNA is extracted from cells of the organism, labelled with a fluorescent dye, and used to bathe the surfaces of the microarrays. As a result of specific base pairing, the RNAs present bind to the genes from which they were originally transcribed and produce fluorescent spots on the chip's surface. Hence, the total set of genes that were transcribed during the biological function of interest can be determined.

Note that forward genetics can aim at a similar goal of assembling the subset of genes that pertain to some specific biological process. The forward genetic approach is to first induce a large set of mutations with phenotypes that appear to change the process in question, and then attempt to define the genes that normally guide the process. However, the technique can only identify genes for which mutations produce an easily recognizable mutant phenotype; genes with subtle effects are often missed.

Comparative genomics

A further application of genomics is in the study of evolutionary relationships. Using classical genetics, evolutionary relationships can be studied by comparing the chromosome size, number, and banding patterns between populations, species, and genera. However, if full genomic sequences are available, comparative genomics brings to bear a resolving power that is much greater than that of classical genetic methods and allows much more subtle differences to be detected. This is because comparative genomics allows the DNAs of organisms to be compared directly and on a small scale. Overall, comparative genomics has shown high levels of similarity between closely related animals, such as humans and chimpanzees, and, more surprisingly, similarity between seemingly distantly related animals, such as humans and insects.

Comparative genomics applied to distinct populations of humans has shown that the human species is a genetic continuum, and the differences between populations are restricted to a very small subset of genes that affect superficial appearance such as skin colour. Furthermore, because DNA sequence can be measured mathematically, genomic analysis can be quantified in a very precise way to measure specific degrees of relatedness. Genomics has detected small-scale changes, such as the existence of surprisingly high levels of gene duplication and mobile elements within genomes.

Nanotechnology

Introduction

Nanoscience concerns a basic understanding of the properties of substances at atomic and near-atomic scales (a nanometre is a billionth of a metre). The term "nanotechnology" is widely used as shorthand to refer to both the science and the technology of this emerging field. Nanotechnology is highly interdisciplinary, involving physics, chemistry, biology, materials science, and the full range of the engineering disciplines. It employs controlled manipulation of physical, chemical, and biological properties to create materials and functional systems with unique capabilities.

Bioassays

One area of intense study in nanomedicine is that of developing new diagnostic tools. Motivation for this work ranges from fundamental biomedical research at the level of single genes or cells to point-of-care applications for health delivery services. With advances in molecular biology, much diagnostic work became focused on detecting specific biological "signatures". These analyses are referred to as bioassays. Examples include studies to determine which genes are active in response to a particular disease or drug therapy. A general approach involves attaching fluorescent dye molecules to the target biomolecules in order to reveal their concentration.

Another type of nanotechnology bioassay involves labelling single-stranded DNA with gold nanoparticles. When the sequence to be detected is present in a solution the gold-labelled molecules interact and the gold particles agglomerate. This causes a large change in optical properties that can be seen in a scanning electron microscope.

Microfluidic systems, or "labs-on-chips", have been developed for biochemical assays of minuscule samples. Typically cramming numerous electronic and mechanical components into a portable unit no larger than a credit card, they are especially useful for conducting rapid analysis in the field.

While these microfluidic systems primarily operate at the microscale (that is, millionths of a metre), nanotechnology has contributed new concepts and will likely play an increasing role in the future. For example, separation of DNA is sensitive to entropic effects, such as the entropy required to unfold DNA of a given length. A new approach to separating DNA could take advantage of its passage through a nanoscale array of posts or channels such that DNA molecules of different lengths would uncoil at different rates. Other researchers have focused on detecting signal changes as nanometre-wide DNA strands are threaded through a nanoscale pore. Early studies used pores punched in membranes by viruses; artificially fabricated nanopores have also been tested. By applying an electric potential across the membrane in a liquid cell to pull the DNA through, changes in ion current can be measured as different repeating base units of the molecule pass through the pores.

Nanotechnology-enabled advances in the entire area of bioassays will clearly impact health care in many ways, from early detection, rapid clinical analysis, and home monitoring to new understanding of molecular biology and genetic-based treatments for fighting disease.

Nanofabrication

Nanotechnology requires new tools for fabrication and measurement. Two very different paths are pursued. One is a top-down strategy of miniaturizing current technologies; the other is a bottom-up strategy of building ever-more-complex molecular devices atom by atom. Top-down approaches are useful for producing structures with long-range order and for making macroscopic connections, whereas bottom-up approaches are best suited for assembly and establishing short-range order at nanoscale dimensions. The integration of top-down and bottom-up techniques is expected to eventually provide the best combination of tools for nanofabrication.

Bottom-up, or self-assembly, approaches to nanofabrication use chemical or physical forces operating at the nanoscale to assemble basic units into larger structures. As component size decreases in nanofabrication, bottom-up approaches provide an increasingly important complement to top-down techniques. A number of bottom-up approaches have been developed for producing nanoparticles, ranging from condensation of atomic vapours on surfaces to coalescence of atoms in liquids. Inspiration for bottom-up approaches comes from biological systems, where nature has harnessed chemical forces to create essentially all the structures needed by life. Researchers hope to replicate nature's ability to produce small clusters of specific atoms, which can then self-assemble into more-elaborate structures.

DNA-assisted assembly may provide a method to integrate hybrid heterogeneous parts into a single device. Biology does this very well, combining self-assembly and self-organization in fluidic environments where weaker electrochemical forces play a significant role. By using DNA-like recognition, molecules on surfaces may be able to direct attachments between objects in fluids. In this approach, polymers made with complementary DNA strands would be used as intelligent "adhesive tape", attaching between polymers only when the right pairing is present. Such assembly might be combined with electrical fields to assist in locating the attachment sites and then be followed by more permanent attachment approaches, such as electrodeposition and metallization. DNA-assisted approaches have several advantages: DNA molecules can be sequenced and replicated in large quantities; DNA sequences act as codes that can be used to recognize complementary DNA strands; hybridized DNA strands form strong bonds to their complementary sequence; and DNA strands can be attached to different devices as labels.

These properties are being explored for ways to self-assemble molecules into nanoscale units. For example, sequences of DNA have been fabricated that adhere only to particular crystal faces of compound semiconductors, providing a basis for self-assembly. By having the correct complementary sequences at the other end of the DNA molecule, certain faces of small semiconductor building blocks can be made that adhere to or repel each other. For example, thiol groups at the end of molecules cause them to attach to gold surfaces, while carboxyl groups can be used for attachment to silica surfaces. Directed assembly is an increasingly important variation of self-assembly where, in quasi-equilibrium environments, parts are moved mechanically, electrically, or magnetically and are placed precisely where they are intended to go.

PART 5

CONTROVERSIES IN GENETICS

10

ETHICAL ISSUES IN GENETICS

Introduction

The human genetic constitution contributes to making people not only what they are – tall or short, male or female, healthy or sick – but also who they are, how they think and feel. Furthermore, although people generally like to think of their genomes as being uniquely theirs, in fact significant aspects are shared with their families, and information about their own genes is also information about their loved ones. Perhaps most important, in the biological sense, genes passed on to children represent the closest humans will ever come to immortality. For these reasons and others, human genetics is a topic fraught with ethical dilemma, with enormous power for good but also with frightening possibilities for misuse. The challenge and responsibility are to harness available information and technologies to improve life and health for all people without compromising privacy, autonomy, or diversity. Of vital importance in achieving these goals is an educated society that is aware of the advantages of new technologies, yet is also concerned about their potential dangers.

Management of Genetic Disease

The management of genetic disease can be divided into counselling, diagnosis, and treatment. In brief, the fundamental purpose of genetic counselling is to help the individual or family understand risks and options and to empower them to make informed decisions. Diagnosis of genetic disease is sometimes clinical, based on the presence of a given set of symptoms, and sometimes molecular, based on the presence of a recognized gene mutation, whether clinical symptoms are present or not. The cooperation of family members may be required to achieve diagnosis for a given individual, and, once accurate diagnosis of that individual has been determined, there may be implications for the diagnoses of other family members. Balancing privacy issues within a family with the ethical need to inform individuals who are at risk for a particular genetic disease can become extremely complex.

Although effective treatments exist for some genetic diseases, for others there are none. It is perhaps this latter set of disorders that raises the most troubling questions with regard to pre-symptomatic testing. Phenotypically healthy individuals can be put in the position of hearing that they are going to become ill and perhaps die prematurely from a genetic disease and that there is nothing they or anyone else can do to stop it. Fortunately, with time and research, this set of disorders is slowly becoming smaller.

Genetic Counselling

Genetic counselling represents the most direct medical application of the advances in understanding of basic genetic mechanisms. Its chief purpose is to help people make responsible and informed decisions concerning their own health or that of their children. Genetic counselling, at least in democratic societies, is non-directive; the counsellor provides information, but decisions are left up to the individual or the family.

Calculating risks of known carriers

Most couples who present themselves for pre-conception counselling fall into one of two categories: those who have already had a child with genetically based problems, and those who have one or more relatives with a disease they think might be inherited. The counsellor must confirm the diagnosis in the affected person with meticulous accuracy, so as to rule out the possibility of alternative explanations for the clinical symptoms observed. A careful family history permits construction of a pedigree that may illuminate the nature of the inheritance (if any), may affect the calculation of risk figures, and may bring to light other genetic influences. The counsellor, a certified health-care professional with special training in medical genetics, must then decide whether the disease in question has a strong genetic component and, if so, whether the heredity is single-gene, chromosomal, or multifactorial.

In the case of single-gene Mendelian inheritance, the disease may be passed on as an autosomal recessive, autosomal dominant, or sex-linked recessive trait, as discussed in the section Classes of Genetic Disease in Chapter 7. If the prospective parents already have a child with an autosomal recessive inherited disease, they both are considered by definition to be carriers, and there is a 25 per cent risk that each future child will be affected. If one of the parents carries a mutation known to cause an autosomal dominant inherited disease, whether that parent is clinically affected or not, there is a 50 per cent risk that each future child will inherit the mutation and therefore may be affected. If, however, the couple has borne a child with an autosomal dominant inherited disease though neither parent carries the mutation, then it will be presumed that a spontaneous mutation has occurred and that there is not a markedly increased risk for recurrence of the disease in future children. There is a caveat to this reasoning, however, because there is also the possibility that the new mutation might have occurred in a progenitor germ cell in one of the parents. Thus, some unknown proportion of that individual's eggs or sperm may carry the mutation, even though the mutation is absent from the somatic cells – including blood, which is generally the tissue sampled for testing. This scenario is called germline mosaicism. Finally, with regard to X-linked disorders, if the pedigree or carrier testing suggests that the mother carries a gene for a sex-linked disease, there is a 50 per cent chance that each son will be affected and that each daughter will be a carrier.

Counselling for chromosomal inheritance most frequently involves either an inquiring couple (consultands) who have had a child with a known chromosomal disorder, such as Down syndrome, or a couple who have experienced multiple miscarriages. To provide the most accurate recurrence risk values to such couples, both parents should be karyotyped to determine if one may be a balanced translocation carrier. Balanced translocations refer to genomic rearrangements in which there is an abnormal covalent arrangement of chromosome segments, although there is no net gain or loss of key genetic material. If both parents exhibit completely normal karyotypes, the recurrence risks cited are low and are strictly empirical.

Most of the common hereditary birth defects, however, are multifactorial. (See the section Diseases Caused by Multifactorial Inheritance in Chapter 7.) If the consulting couple have had one affected child, the empirical risk for each future child will be about 3 per cent. If they have borne two affected children, the chance of recurrence will rise to about 10 per cent. Clearly these are population estimates, so that the risks within individual families may vary.

Estimating probability: Bayes's theorem

As described above, the calculation of risks is relatively straightforward when the consultands are known carriers of diseases due to single genes of major effect that show regular Mendelian inheritance. For a variety of reasons, however, the parental genotypes frequently are not clear and must be approximated from the available family data. Bayes's theorem, a statistical method first devised by the British clergyman and scientist Thomas Bayes in 1763, can be used to assess the relative probability of two or more alternative possibilities (e.g. whether a consultand is or is not a carrier). The likelihood derived from the appropriate Mendelian law (prior probability) is combined with any additional information that has been obtained from the consultand's family history or from any tests performed (conditional probability). A joint probability is then determined for each alternative outcome by multiplying the prior probability by all conditional probabilities. By dividing the joint probability of each alternative by the sum of all joint probabilities, the posterior probability is arrived at. Posterior probability is the likelihood that the individual, whose genotype is uncertain, either carries the mutant gene or does not. One example application of this method, applied to the sex-linked recessive disease Duchenne muscular dystrophy (DMD), is given below.

In this example, the consultand wishes to know her risk of having a child with DMD. The family's pedigree is illustrated in Figure 10.1. It is known that the consultand's grandmother (I-2) is a carrier, since she had two affected sons (spontaneous mutations occurring in both brothers would be extremely unlikely). What is uncertain is whether the consultand's mother (II-4) is also a carrier.

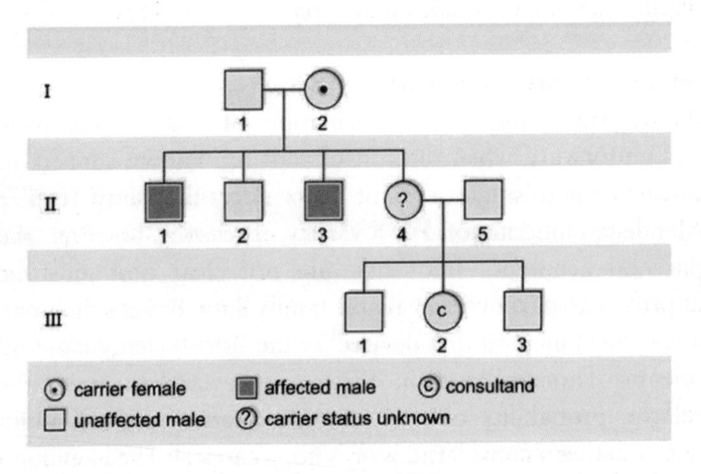

Figure 10.1 Pedigree of a family with a history of Duchenne muscular dystrophy, which is carried by females (circles) and affects half of a carrier's male children (squares).

The Bayesian method for calculating the consultand's risk is as follows:

If II-4 is a carrier (risk = 1/5), then there is a 1/2 chance that the consultand is also a carrier, so her total empirical risk is 1/5 × 1/2 = 1/10. If she becomes pregnant, there is a 1/2 chance that her child will be male and a 1/2 chance that the child, regardless of sex, will inherit the familial mutation. Hence, the total empirical risk for the consultand (III-2) to have an affected child is 1/10 × 1/2 × 1/2 = 1/40. Of course, if the familial mutation is known, presumably from molecular testing of an affected family member, the carrier status of III-2 could be determined directly by molecular analysis, rather than estimated by Bayesian calculation. If the family is cooperative and an affected member is available for study, this is clearly the most informative route to follow, because the risk for the consultand to carry the familial mutation would be either 1 or 0, and not 1/10. If her risk is 1, then each of her sons will have a 1/2 chance of being affected. If her risk is 0, none of her children will be affected (unless a new mutation occurs, which is very rare).

After determining the nature of the heredity, the counsellor discusses with the consultand the likely risks and the available options to minimize impact of those risks on the individual and the family. In the case of a couple in which one member has a family history of a genetic disorder – for example, cystic fibrosis – typical options might include any of the following choices: (1) Accept the risks and take a chance that any future children may be affected. (2) Seek molecular testing for known mutations of cystic fibrosis in relevant family members to determine with greater accuracy whether either or both prospective parents are carriers for this recessive disorder. (3) If both members of the couple are found to be carriers, utilize donor sperm for artificial insemination. This option is a good genetic solution only if the husband carries a dominant mutation, or if both parents are carriers of a recessive mutation. If the recessive trait is reasonably common, as are mutations for cystic fibrosis, however, it would be reasonable to ask that the sperm donor be checked for carrier status before pursuing this option. (4) Proceed with natural reproduction, but pursue prenatal diagnosis with the possibility of selective termination of an affected pregnancy, if desired by the parents. (5) Pursue in vitro fertilization with donor eggs, if the woman is the at-risk partner, or use both eggs and sperm from the couple but employ preimplantation diagnostics to select only unaffected embryos for implantation (see below). (6) Decide against biological reproduction because the risks and available options are unacceptable; possibly pursue adoption.

Diagnosis

Prenatal diagnosis

Perhaps one of the most sensitive areas of medical genetics is prenatal diagnosis, the genetic testing of an unborn fetus, because of fears of eugenic misuse or because some couples may choose to terminate a pregnancy depending on the outcome of the test. Nonetheless, prenatal testing in one form or another is ubiquitous in most industrialized countries, and recent advances, both in testing technologies and in the set of "risk factor" genes to be screened, promise to make prenatal diagnosis even more widespread. Indeed, parents may soon be able to ascertain information not only about the sex and health status of their unborn child but also about his or her complexion, personality, and intellect. Whether parents should have access to all of this information and how they may choose to use it are matters of much debate.

Current forms of prenatal diagnosis can be divided into two classes, those that are apparently non-invasive and those that are more invasive. At present the non-invasive tests are generally offered to all pregnant women, whereas the more invasive tests are generally recommended only if some risk factors exist. The non-invasive tests include ultrasound imaging and maternal serum tests. Serum tests include one for alphafetoprotein (AFP) or one for AFP, oestriol, and human chorionic gonadotropin (triple screen). These tests serve as screens for structural fetal malformations and for neural tube closure defects. The triple screen also can detect some cases of Down syndrome, although there is a significant false-positive and false-negative rate.

More invasive tests include amniocentesis, chorionic villus sampling, percutaneous umbilical blood sampling, and, on rare occasions, preimplantation testing of either a polar body or a dissected embryonic cell. Amniocentesis is a procedure in which a long, thin needle is inserted through the abdomen and uterus into the amniotic sac, enabling the removal of a small amount of the amniotic fluid bathing the fetus. This procedure is generally performed under ultrasound guidance between the 15th and 17th weeks of pregnancy, and, although it is generally regarded as safe, complications can occur, ranging from cramping to infection or loss of the fetus. The amniotic fluid obtained can be used in three ways: (1) Living fetal cells recovered from this fluid can be induced to grow and can be analysed to assess chromosome number, composition, or structure. (2) Cells recovered from the fluid can be used for molecular studies. (3) The amniotic fluid itself can be analysed biochemically to determine the relative abundance of a variety of compounds associated with normal or abnormal fetal metabolism and development. Amniocentesis is typically offered to pregnant women over age 35, because of the significantly increased rate of chromosome disorders observed in the children of older mothers. A clear advantage of amniocentesis is the wealth of material obtained and the relative safety of the procedure. The disadvantage is timing: results may not be received until the pregnancy is already into the 19th week or beyond, at which point the possibility of termination may be much more physically and emotionally wrenching than if considered earlier.

Chorionic villus sampling (CVS) is a procedure in which either a needle is inserted through the abdomen or a thin tube is inserted into the vagina and cervix to obtain a small sample of placental tissue called chorionic villi. CVS has the advantage of being performed earlier in the pregnancy (generally 10–11 weeks), although the risk of complications is greater than that for amniocentesis. Risks associated with CVS include fetal loss and fetal limb reduction if the procedure is performed earlier than 10 weeks gestation. Another disadvantage of CVS reflects the tissue sampled: chorionic villi are not part of the embryo, and such a sample may not accurately represent the embryonic genetic constitution. In contrast, amniotic cells are embryonic in origin, having been sloughed off into the fluid. Therefore, abnormalities, often chromosomal, may be seen in the chorionic villi but not in the fetus, or vice versa.

Both percutaneous umbilical blood sampling (PUBS) and preimplantation testing are rare, relatively high-risk, and performed only in very unusual cases. Preimplantation testing of embryos derived by in vitro fertilization is a particularly new technique and is currently used only in cases of couples who are at high risk for having a fetus affected with a given familial genetic disorder and who find all other alternatives unacceptable. Preimplantation testing involves obtaining eggs and sperm from the couple, combining them in the laboratory, and allowing the resultant embryos to grow until they reach the early blastocyst stage of development, at which point a single cell is removed from the rest and harvested for fluorescent in situ hybridization (FISH) or molecular analysis. The problem with this procedure is that one cell is scant material for diagnosis, so that a large array of tests cannot be performed. Similarly, if the test fails for any technical reason, it cannot be repeated. Finally, embryos that are found to be normal and therefore selected for implantation into the mother are subject to other complications normally associated with in vitro fertilization – namely, that only a small fraction of the implanted embryos make it to term and that multiple, and therefore high-risk, pregnancies are common. Nonetheless, many at-risk couples find these complications easier to accept than the elective termination of the pregnancy.

It should be noted that researchers have identified fetal cells in the maternal circulation and that procedures are under development to enable their isolation and analysis, thereby providing a non-invasive alternative for molecular prenatal testing. Although these techniques are currently experimental and are not yet available for clinical application, they may well become the methods of choice in the future.

Genetic testing

In the case of genetic disease, options often exist for presymptomatic diagnosis – that is, diagnosis of individuals at risk for developing a given disorder, even though at the time of diagnosis they may be clinically healthy. Options may even exist for carrier testing, studies that determine whether an individual is at increased risk of having a child with a given disorder, even though he or she personally may never display symptoms. Accurate predictive information can enable early intervention, which often prevents the clinical onset of symptoms and the irreversible damage that may have already occurred by waiting for symptoms and then responding to them. In the case of carrier testing, accurate information can enable prospective parents to make more-informed family-planning decisions. Unfortunately, there can also be negative aspects to early detection, including such issues as privacy, individual responses to potentially negative information, discrimination in the workplace, or discrimination in access to or cost of health or life insurance. While some governments have outlawed the use of pre-symptomatic genetic testing information by insurance companies and employers, others have embraced it as a way to bring spiralling health-care costs under control. Some communities have even considered instituting premarital carrier testing for common disorders in the populace.

Genetic testing procedures can be divided into two different groups: (1) testing of individuals considered at risk from phenotype or family history; and (2) screening of entire populations, regardless of phenotype or personal family history, for evidence of genetic disorders common in that population. Both forms are currently pursued in many societies. Indeed, with the explosion of information about the human genome and the increasing identification of potential "risk genes" for common disorders, such as cancer, heart disease, or diabetes, the role of predictive genetic screening in general medical practice is likely to increase.

Adults are generally tested for evidence of genetic disease only if personal or family history suggests they are at increased risk for a given disorder. A typical example would be a young man whose father, paternal aunt, and older brother have all been diagnosed with early-onset colorectal cancer. Although this person may appear perfectly healthy, he is at significantly increased risk of carrying mutations associated with familial colorectal cancer, and accurate genetic testing could enable heightened surveillance (e.g. frequent colonoscopies) that might ultimately save his life.

Carrier testing for adults in most developed countries is generally offered only if family history or ethnic origins suggest an increased risk of having a particular disease. A typical example would be to offer carrier testing for cystic fibrosis to a couple including one member who has a sibling with the disorder. Another would be to offer carrier testing for Tay–Sachs disease to couples of Ashkenazic Jewish origin, a population known to carry an increased frequency of Tay–Sachs mutations. The same would be true for couples of African or

Mediterranean descent with regard to sickle cell anaemia or thalassaemia, respectively. Typically, in each of these cases a genetic counsellor would be involved to help the individuals or couples understand their options and make informed decisions.

Screening of large unphenotyped populations for evidence of genetic disease is currently pursued in most industrialized countries only in the newborn population, although future developments in the identification of risk genes for common adult-onset disorders may change this policy. So-called mandated newborn screening was initiated in many societies in the latter quarter of the 20th century in an effort to prevent the drastic and often irreversible damage associated with a small number of relatively common genetic disorders whose sequelae can be either prevented or significantly relieved by early detection and intervention. The general practice is to collect a small sample of blood from each newborn, generally by pricking the infant's heel and collecting drops of blood on special filter paper, which is then analysed. Perhaps the best-known disorder screened in this manner is phenylketonuria (PKU), an autosomal recessive inborn error of metabolism discussed in the section Autosomal recessive inheritance in Chapter 7. With early diagnosis and dietary intervention that is maintained throughout life, children with PKU can escape intellectual disability and grow into healthy adults who lead full and productive lives.

Although many of the genetic disorders tested by mandated newborn screening are metabolic in nature, this trend is changing. For example, in some communities newborns are screened for profound congenital hearing loss, which is known to be frequently genetic in origin and for which effective intervention is available (e.g. through cochlear implants).

Genetic tests themselves can take many forms, and the choice of tests depends on a number of factors. For example, screening for evidence of sickle cell anaemia, a haemoglobin disorder, is generally pursued at least initially by tests involving the haemoglobin proteins themselves, rather than DNA, because the relevant gene product (blood) is readily accessible, and because the protein test is currently cheaper to perform than the DNA test. In contrast, screening for cystic fibrosis, a disorder that predominantly affects the lungs and pancreas, is generally pursued in the at-risk newborn at the level of DNA because there is no cheap and accurate alternative. Older persons suspected of having cystic fibrosis, however, can also be diagnosed with a "sweat test" that measures sweat electrolytes.

Tests involving analysis of DNA are particularly powerful because they can be performed using very tiny samples; also, the DNA tested can originate from almost any tissue type, regardless of whether the gene of interest happens to be expressed in that tissue. Current technologies applied for mutation detection include traditional karyotyping and Southern blotting, as well as a multitude of other tests, including FISH with specific probes or PCR. Which tests are applied depends on whether the genetic abnormalities are likely to be: (1) chromosomal (in which case karyotyping or FISH are appropriate); (2) large deletions or other rearrangements (best tested for by Southern blotting or PCR); or (3) point mutations (best confirmed by PCR followed by oligonucleotide hybridization or restriction enzyme digestion). If a large number of different point mutations are sought, as is often the case, the most appropriate technology may be microarray hybridization analysis, which can test for tens to hundreds of thousands of different point mutations in the same sample simultaneously.

Options for Treatment

The options for the treatment of genetic disease are both many and expanding. Although a significant number of genetic diseases still have no effective treatment, for many the treatments are quite good. Current approaches include dietary management, such as the restriction of phenylalanine in PKU; protein or enzyme replacement, such as that used in Gaucher syndrome, haemophilia, and diabetes; and tissue replacement, such as blood transfusions or bone marrow transplantation in sickle cell anaemia and thalassaemia. Other treatments are strictly symptomatic, such as the use of splints in Ehlers–Danlos syndrome, administration of antibiotics in early cystic fibrosis, or female hormone replacement in Turner syndrome. Many options involve surveillance and surgery, such as regular checks of aortic root diameter followed by surgery to prevent aortic dissection in Marfan syndrome, or regular colonoscopies in persons at risk for familial colon cancer followed by surgical removal of the colon at the first signs of disease.

Some genetic diseases may also be amenable to treatment by gene therapy, the introduction of normal genetic sequences to replace or augment the inherited gene whose mutation underlies the disease. Although some successes have been reported with gene therapy trials in humans – for example, with patients who have severe combined immunodeficiency (SCID) or haemophilia – significant technical challenges remain.

Race

The Decline of "Race" in Science

Mendelian heredity and the development of blood group systems

In 1900, after the rediscovery of Gregor Mendel's experiments dealing with heredity, scientists began to focus greater attention on genes and chromosomes. Their objective was to ascertain the hereditary basis for numerous physical traits. Once the ABO blood group system was discovered and was shown to follow the pattern of Mendelian heredity, other systems – the MN system, the rhesus system, and many others – soon followed (see Chapter 6, Blood Types). Experts thought that at last they had found genetic features that, because they are inherited and not susceptible to environmental influences, could be used to identify races. By the 1960s and 1970s, scientists were writing about racial groups as populations that differed from one another not in absolute features but in the frequencies of expression of genes that all populations share. It was expected that each race, and each population within each race, would have frequencies of certain ascertainable genes that would mark them off from other races.

Information on blood groups was taken from large numbers of populations, but when scientists tried to show a correlation of blood group patterns with the conventional races, they found none. While populations differed in their blood group patterns, in such features as the frequencies of A, B, and O types, no evidence was found to document race distinctions. As knowledge of human heredity expanded, other genetic markers of difference were sought, but these also failed to neatly separate humanity into races. Most differences are expressed in subtle gradations over wide geographic space, not in abrupt changes from one "race" to another. Moreover, not all groups within a large "geographic race" share the same patterns of genetic features. The internal variations within races have proved to be greater than those between races. Most importantly, physical, or phenotypic, features determined by DNA are inherited independently of one another, further frustrating attempts to describe race differences in genetic terms.

"Race" and Intelligence

Anthropometric measurements did not provide any data to prove group superiority or inferiority. As various fields of study emerged in the late 19th century, some scholars began to focus on mental traits as a means to examine and describe human differences. Psychology as a growing field began developing its own programmatic interests in discovering race differences.

In the 1890s the psychologist Alfred Binet began testing the mental abilities of French schoolchildren to ascertain how children learned and to help those who had trouble learning. Binet did not call his test an intelligence test, and its purpose was not to divide French schoolchildren into hierarchical groups. But with these tests, a new mechanism was born that would provide powerful support to those who held beliefs in racial differences in intelligence.

Psychologists in the United States very quickly adopted Binet's tests and modified them for American use. More than that, they reinterpreted the results to be clear evidence of innate intelligence. Lewis Terman and his colleagues at Stanford University developed the Stanford–Binet IQ (intelligence quotient) test, which set the standard for similar tests produced by other American psychologists.

IQ tests began to be administered in great number during the second decade of the 20th century. The influences of hereditarian beliefs and the power of the racial world view had conditioned Americans to believe that intelligence was inherited and permanent and that no external influences could affect it. Indeed, heredity was thought to determine a person's or a people's place in life and success or failure. IQ tests came to be employed more in the United States than in any other country. A major reason for this was that the tests tended to confirm the expectations of white Americans; on average, black Americans did less well than whites on IQ tests. But the tests also revealed that the disadvantaged of all races do worse on IQ tests than do the privileged. Such findings were compatible with the beliefs of large numbers of Americans who had come to accept unqualified biological determinism.

Opponents of IQ tests and their interpretations argued that intelligence had not been clearly defined, that experts did not agree on its definition, and that there were many different types of intelligence that cannot be measured. They also called attention to the many discrepancies and contradictions of the tests. One of the first examples of empirical evidence against the "innate intelligence" arguments was the revelation by psychologist Otto Klineberg in the 1930s that blacks in four northern states did better on average than whites in the four southern states where expenditures on education were lowest. Klineberg's analysis pointed to a direct correlation between income and social class and performance on IQ tests. Further evidence indicated that students with the best primary education and greater cultural experiences always did better on such tests. Experts thus argued that such tests are culture-bound – that is, they reflect and measure the cultural experiences and knowledge of those who take the tests and their levels of education and training.

Heritability and Malleability of Intelligence

Intelligence has historically been conceptualized as a more or less fixed trait. However, a minority of investigators believe either that it is highly heritable or that it is minimally heritable, but most take an intermediate position.

Among the most fruitful methods that have been used to assess the heritability of intelligence is the study of identical twins who were separated at an early age and reared apart. If the twins were raised in separate environments, and if it is assumed that when twins are separated they are randomly distributed across environments (often a dubious assumption), then the twins would have in common all of their genes but none of their environment, except for chance environmental overlap. As a result, the correlation between their performance on intelligence tests could identify any possible link between test scores and heredity. Another method compares the relationship between intelligence-test scores of identical twins and those of fraternal twins. Because these results are computed on the basis of intelligence-test scores, however, they represent only those aspects of intelligence that are measured by the tests.

Studies of twins do in fact provide strong evidence for the heritability of intelligence; the scores of identical twins reared apart are highly correlated. In addition, adopted children's scores are highly correlated with the scores of their birth parents and not with the scores of their adoptive parents. Also significant are findings that heritability can differ between ethnic and racial groups, as well as across time within a single group; that is, the extent to which genes versus environment matter in IQ depends on many factors, including socio-economic class. Moreover, the American behaviour geneticist Robert Plomin and others have found that evidence of the heritability of intelligence increases with age; this suggests that, as a person grows older, genetic factors become a more important determinant of intelligence, while environmental factors become less important.

Whatever the heritability factor of IQ may be, whether intelligence can be increased is a separate issue. Evidence that it can was provided by the American-born New Zealand political scientist James Flynn, who showed that intelligence test scores around the world rose steadily in the late 20th century. The reasons for the increase are not fully understood, however, and the phenomenon thus requires additional careful investigation. For example, among many possible causes of the increase are environmental changes such as the addition of vitamin C to prenatal and postnatal diet and, more generally, the improved nutrition of mothers and infants as compared with earlier in the century. In their book *The Bell Curve* (1994), American psychologist Richard Herrnstein and American social scientist Charles Murray argued that IQ is important for life success and that differences between racial groups in life success can be attributed in part to differences in IQ. They speculated that these differences might be genetic. As noted above, such claims remain speculative.

Despite the general increase in scores, average IQs continue to vary both across countries and across different socio-economic groups. For example, many researchers have found a positive correlation between socio-economic status and IQ, although they disagree about the reasons for the relationship. Most investigators also agree that differences in educational opportunities play an important role, though some believe that the main basis of the difference is hereditary. There is no broad agreement about why such differences exist. Most important, it should be noted that these differences are based on IQ alone and not on intelligence as it is more broadly defined. Even less is known about group differences in intelligence as it is broadly defined than is known about differences in IQ. Nevertheless, theories of inherited differences in IQ between racial groups have been found to be without basis. There is more variability within groups than between groups.

Finally, no matter how heritable intelligence may be, some aspects of it are still malleable. With intervention, even a highly heritable trait can be modified. A programme of training in intellectual skills can increase some aspects of a person's intelligence; however, no training programme – no environmental condition of any sort – can make a genius of a person with low measured intelligence. But some gains are possible, and programmes have been developed for increasing intellectual skills. Intelligence, in the view of many authorities, is not a foregone conclusion the day a person is born. A main trend for psychologists in the intelligence field has been to combine testing and training functions to help people make the most of their intelligence.

Stem Cells

A stem cell is an undifferentiated cell that can divide to produce some offspring cells that continue as stem cells and some cells that are destined to differentiate (become specialized). Stem cells are an ongoing source of the differentiated cells that make up the tissues and organs of animals and plants. There is great interest in stem cells because they have potential in the development of therapies for replacing defective or damaged cells resulting from a variety of disorders and injuries, such as Parkinson disease, heart disease, and diabetes. There are two major types of stem cells: embryonic stem cells and adult stem cells, which are also called tissue stem cells.

Embryonic Stem Cells

Embryonic stem cells are stem cells that are derived from the inner cell mass of a mammalian embryo at a very early stage of development, when it is composed of a hollow sphere of dividing cells (a blastocyst). Embryonic stem cells from human embryos and from embryos of certain other mammalian species can be grown in tissue culture.

Mouse embryonic stem cells

The most studied embryonic stem cells are mouse embryonic stem cells, which were first reported in 1981. This type of stem cell can be cultured indefinitely in the presence of leukaemia inhibitory factor (LIF), a glycoprotein cytokine. If cultured mouse embryonic stem cells are injected into an early mouse embryo at the blastocyst stage, they will become integrated into the embryo and produce cells that differentiate into most or all of the tissue types that subsequently develop. This ability to repopulate mouse embryos is the key defining feature of embryonic stem cells, and because of it they are considered to be pluripotent – that is, able to give rise to any cell type of the adult organism. If the embryonic stem cells are kept in culture in the absence of LIF, they will differentiate into "embryoid bodies", which somewhat resemble early mouse embryos at the egg-cylinder stage, with embryonic stem cells inside an outer layer of endoderm. If embryonic stem cells are grafted into an adult mouse, they will develop into a type of tumour called a teratoma, which contains a variety of differentiated tissue types.

Mouse embryonic stem cells are widely used to create genetically modified mice. This is done by introducing new genes into embryonic stem cells in tissue culture, selecting the particular genetic variant that is desired, and then inserting the genetically modified cells into mouse embryos. The resulting "chimeric" mice are composed partly of host cells and partly of the donor embryonic stem cells. As long as some of the chimeric mice have germ cells (sperm or eggs) that have been derived from the embryonic stem cells, it is possible to breed a line of mice that have the same genetic constitution as the embryonic stem cells and therefore incorporate the genetic modification that was made in vitro. This method has been used to produce thousands of new genetic lines of mice. In many such genetic lines, individual genes have been ablated in order to study their biological function; in others, genes have been introduced that have the same mutations that are found in various human genetic diseases. These "mouse models" for human disease are used in research to investigate both the pathology of the disease and new methods for therapy.

Human embryonic stem cells

Extensive experience with mouse embryonic stem cells made it possible for scientists to grow human embryonic stem cells from early human embryos, and the first human stem cell line was created in 1998. Human embryonic stem cells are in many respects similar to mouse embryonic stem cells, but they do not require LIF for their maintenance. The human embryonic stem cells form a wide variety of differentiated tissues in vitro, and they form teratomas when grafted into immunocompetent mice. It is not known whether the cells can colonize all the tissues of a human embryo, but it is presumed from their other properties that they are indeed pluripotent cells, and they are therefore regarded as a possible source of differentiated cells for cell therapy – the replacement of a patient's defective cell type with healthy cells. Large quantities of cells, such as dopamine-secreting neurons for the treatment of Parkinson disease and insulin-secreting pancreatic beta cells for the treatment of diabetes, could be produced from embryonic stem cells for cell transplantation. Cells for this purpose have previously been obtainable only from sources in very limited supply, such as the pancreatic beta cells obtained from the cadavers of human organ donors.

The use of human embryonic stem cells evokes ethical concerns, because the blastocyst-stage embryos are destroyed in the process of obtaining the stem cells. The embryos from which stem cells have been obtained are produced through in vitro fertilization, and people who consider preimplantation human embryos to be human beings generally believe that such work is morally wrong. Others accept it because they regard the blastocysts to be simply balls of cells, and human cells used in laboratories have not previously been accorded any special moral or legal status. Moreover, it is known that none of the cells of the inner cell mass are exclusively destined to become part of the embryo itself – all of the cells contribute some or all of their cell offspring to the placenta, which also has not been accorded any special legal status. The divergence of views on this issue is illustrated by the fact that the use of human embryonic stem cells is allowed in some countries and prohibited in others.

Embryonic germ cells

Embryonic germ cells, derived from primordial germ cells found in the gonadal ridge of a late embryo, have many of the properties of embryonic stem cells. The primordial germ cells in an embryo develop into stem cells that in an adult generate the reproductive gametes (sperm or eggs). In mice and humans it is possible to grow embryonic germ cells in tissue culture with the appropriate growth factors – namely, LIF and another cytokine called fibroblast growth factor.

Adult Stem Cells

Some tissues in the adult body, such as the epidermis of the skin, the lining of the small intestine, and bone marrow, undergo continuous cellular turnover. They contain stem cells, which persist indefinitely, and a much larger number of "transit amplifying cells", which arise from the stem cells and divide a finite number of times until they become differentiated. The stem cells exist in niches formed by other cells, which secrete substances that keep the stem cells alive and active. Some types of tissue, such as liver tissue, show minimal cell division or undergo cell division only when injured. In such tissues there is probably no special stem-cell population, and any cell can participate in tissue regeneration when required.

Epithelial stem cells

The epidermis of the skin contains layers of cells called keratinocytes. Only the basal layer, next to the dermis, contains cells that divide. A number of these cells are stem cells, but the majority are transit amplifying cells. The keratinocytes slowly move outward through the epidermis as they mature, and they eventually die and are sloughed off at the surface of the skin. The epithelium of the small intestine forms projections called villi, which are interspersed with small pits called crypts. The dividing cells are located in the crypts, with the stem cells lying near the base of each crypt. Cells are continuously produced in the crypts, migrate on to the villi, and are eventually shed into the lumen of the intestine. As they migrate, they differentiate into the cell types characteristic of the intestinal epithelium.

Bone marrow and haematopoietic stem cells

Bone marrow contains cells called haematopoietic stem cells, which generate all the cell types of the blood and the immune system. Haematopoietic stem cells are also found in small numbers in peripheral blood and in larger numbers in umbilical cord blood. In bone marrow, haematopoietic stem cells are anchored to osteoblasts of the trabecular bone and to blood vessels. They generate progeny that can become lymphocytes, granulocytes, red blood cells, and certain other cell types, depending on the balance of growth factors in their immediate environment.

Work with experimental animals has shown that transplants of haematopoietic stem cells can occasionally colonize other tissues, with the transplanted cells becoming neurons, muscle cells, or epithelia. The degree to which transplanted haematopoietic stem cells are able to colonize other tissues is exceedingly small. Despite this, the use of haematopoietic stem cell transplants is being explored for conditions such as heart disease or autoimmune disorders. It is an especially attractive option for those opposed to the use of embryonic stem cells.

Bone marrow transplants (also known as bone marrow grafts) represent a type of stem cell therapy that is in common use. They are used to allow cancer patients to survive otherwise lethal doses of radiation therapy or chemotherapy that destroy the stem cells in bone marrow. For this procedure, the patient's own marrow is harvested before the cancer treatment and is then re-infused into the body after treatment. The haematopoietic stem cells of the transplant colonize the damaged marrow and eventually repopulate the blood and the immune system with functional cells. Bone marrow transplants are also often carried out between individuals (allografts). In this case the grafted marrow has some beneficial anti-tumour effect. Risks associated with bone marrow allografts include rejection of the graft by the patient's immune system and reaction of immune cells of the graft against the patient's tissues (graft-versus-host disease).

Neural stem cells

Research has shown that there are also stem cells in the brain. In mammals very few new neurons are formed after birth, but some neurons in the olfactory bulbs and in the hippocampus are continually being formed. These neurons arise from neural stem cells, which can be cultured in vitro in the form of neurospheres – small cell clusters that contain stem cells and some of their progeny. This type of stem cell is being studied for use in cell therapy to treat Parkinson disease and other forms of neurodegeneration or traumatic damage to the central nervous system.

Somatic Cell Nuclear Transfer

Following experiments in animals, including those used to create Dolly the sheep (see below), there has been much discussion about the use of somatic cell nuclear transfer (SCNT) to create pluripotent human cells. In SCNT the nucleus of a somatic cell (a fully differentiated cell, excluding germ cells), which contains the majority of the cell's DNA, is removed and transferred into an unfertilized egg cell that has had its own nuclear DNA removed. The egg cell is grown in culture until it reaches the blastocyst stage. The inner cell mass is then removed from the egg, and the cells are grown in culture to form an embryonic stem cell line (generations of cells originating from the same group of parent cells). In theory, these cells could be stimulated to differentiate into various types of cells needed for transplantation. Since these cells would be genetically identical to the original donor, they could be used to treat the donor with no problems of immune rejection.

In 2008 SCNT was successfully achieved with human cells for the first time. However, the procedure is very controversial for several reasons. One is that SCNT is inefficient, sometimes requiring dozens of eggs before one egg successfully produces embryonic stem cells. Human eggs are in short supply, and there are many legal and ethical problems associated with egg donation. There are also unknown risks involved with transplanting SCNT-derived stem cells into humans, because the mechanism by which the unfertilized egg is able to reprogram the nuclear DNA of a differentiated cell is not entirely understood. In addition, SCNT is commonly used to produce clones of animals (such as Dolly). Although the cloning of humans is currently illegal throughout the world, the fertilized egg cell that contains nuclear DNA from an adult cell could in theory be re-implanted into a woman's uterus and come to term as an actual cloned human. Thus, there exists strong opposition among some groups to the use of SCNT to generate human embryonic stem cells.

Induced Pluripotent Stem Cells

Because of the ethical and moral issues surrounding the use of embryonic stem cells, scientists have searched for ways to reprogram adult somatic cells. Studies of cell fusion, in which differentiated adult somatic cells grown in culture with embryonic stem cells fuse with the stem cells and acquire properties like those of embryonic stem cells, led to the idea that specific genes could reprogram differentiated adult cells. An advantage of cell fusion is that it relies on existing embryonic stem cells instead of eggs. However, fused cells stimulate an immune response when transplanted into humans, which leads to transplant rejection. As a result, research has become increasingly focused on the genes and proteins capable of reprogramming adult cells to a pluripotent state. In order to make adult cells pluripotent without fusing them to embryonic stem cells, regulatory genes that induce pluripotency must be introduced into the nuclei of adult cells. To do this, adult cells are grown in cell culture, and specific combinations of regulatory genes are inserted into retroviruses (viruses that convert RNA into DNA), which are then introduced to the culture medium. The retroviruses transport the RNA of the regulatory genes into the nuclei of the adult cells, where the genes are then incorporated into the DNA of the cells. About 1 out of every 10,000 cells acquires embryonic stem cell properties. Although the mechanism is still uncertain, it is clear that some of the genes confer embryonic stem cell properties by means of the regulation of numerous other genes. Adult cells that become reprogrammed in this way are known as induced pluripotent stem cells (iPS).

Similar to embryonic stem cells, iPS cells can be stimulated to differentiate into select types of cells that could in principle be used for disease-specific treatments. In addition, as was originally proposed for SCNT, induced pluripotent stem cells could be used to generate patient-specific stem cells, thus avoiding cell rejection by the immune system. However, before induced pluripotent stem cells can be used to treat human diseases, researchers must find a way to introduce the active reprogramming genes without using retroviruses, which can cause diseases such as leukaemia in humans.

Genetically Modified Organisms

Organisms whose genomes have been altered in order to favour the expression of desired physiological traits or the production of desired biological products are called genetically modified organisms (GMOs). In conventional livestock production, crop farming, and even pet breeding, it has long been a practice to breed select individuals of a species in order to arrive at offspring that have desirable traits. In genetic modification, sophisticated genetic technologies are employed to arrive at organisms whose genomes have been precisely designed at the molecular level, usually by the inclusion of genes from unrelated species of organisms that code for traits that would not be obtainable through conventional selective breeding.

Modern GMOs are produced using scientific methods that include recombinant DNA technology and cloning. Cloning technology generates offspring that are genetically identical to the parent by the transfer of an entire donor nucleus into the enucleated cytoplasm of a host egg. The first animal produced using this cloning technique was a sheep, named Dolly, born in 1996. Since then, a number of other animals, including pigs, horses, and dogs, have been generated using reproductive cloning technology. Recombinant DNA technology, on the other hand, involves the insertion of one or more individual genes from an organism of one species into the DNA of another. Whole-genome replacement, involving the transplantation of one bacterial genome into the "body", or cytoplasm, of another microorganism, has been reported, although this technology is still limited to basic scientific applications.

GMOs produced through modern genetic technologies have become a part of everyday life, interacting with society through agriculture, medicine and research, and environmental management. However, while GMOs have benefited human society in many ways, some disadvantages exist, making the production of GMOs highly controversial worldwide.

GMOs in Agriculture

Genetically modified (GM) foods were first approved for human consumption in the United States in 1995, and by 1999 almost 50 per cent of the corn, cotton, and soybeans planted in the United States were GM. The introduction of these crops has dramatically increased per area crop yields and, in some cases, has reduced the use of insecticides. For example, the application of wide-spectrum insecticides has declined in many areas growing plants, such as potatoes, cotton, and corn, that have been endowed by a gene from the bacterium *Bacillus thuringiensis* that produces a natural insecticide called Bt toxin. Field studies conducted in India in which Bt cotton was compared to non-Bt cotton demonstrated a 30–80 per cent increase in yield from the GM crop. This increase was attributed to marked improvement in the GM plants' ability to overcome bollworm infestation.

By 2002 more than 60 per cent of processed foods consumed in the United States contained at least some GM ingredients. Despite the concerns of some consumer groups, especially in Europe, numerous scientific panels, including the U.S. Food and Drug Administration, have concluded that consumption of GM foods is safe, even in cases involving GM foods with genetic material from very distantly related organisms. By 2006, although the majority of GM crops were still grown in the Americas, GM plants tailored for production and consumption in other parts of the world were in field tests. For example, sweet potatoes intended for Africa were modified for resistance to sweet potato feathery mottle virus (SPFMV) by inserting into the sweet potato genome a gene encoding a viral coat protein from the strain of virus that causes SPFMV. The premise for this modification was based on earlier studies in other plants, such as tobacco, in which introduction of viral coat proteins rendered plants resistant to the virus. In addition, so-called "golden" rice intended for Asia was modified to produce almost 20 times the beta-carotene of previous varieties. Golden rice was created by modifying the rice genome to include a gene from the daffodil *Narcissus pseudonarcissus* that produces an enzyme known as phyotene synthase and a gene from the bacterium *Erwinia uredovora* that produces an enzyme called phyotene desaturase. The introduction of these genes enabled beta-carotene, which is converted to vitamin A in the human liver, to accumulate in the endosperm – the edible part – of the rice plant, thereby increasing the amount of beta-carotene available for vitamin A synthesis in the body. Another form of modified rice was generated to help combat iron deficiency, which impacts close to 30 per cent of the world population. This GM crop was engineered by introducing into the rice genome a ferritin gene from the common bean, *Phaseolus vulgaris*, that produces a protein capable of binding iron, as well as a gene from the fungus *Aspergillus fumigatus* that produces an enzyme capable of digesting compounds that increase iron bioavailability via digestion of phytate (an inhibitor of iron absorption). In addition, the iron-fortified GM rice was engineered to overexpress an existing rice gene that produces a cysteine-rich metal-binding protein that enhances iron absorption. A variety of other crops modified to endure the weather extremes common in other parts of the globe are also in production.

GMOs in Medicine and Research

GMOs have emerged as one of the mainstays of biomedical research since the 1980s. For example, GM animal models of human genetic diseases have enabled researchers to test novel therapies and to explore the roles of candidate risk factors and modifiers of disease outcome. GM microbes, plants, and animals have also revolutionized the production of complex pharmaceuticals by enabling the generation of safer and cheaper vaccines and therapeutics. Pharmaceutical products include recombinant hepatitis B vaccine produced by GM baker's yeast; injectable insulin (for diabetics) produced in GM *Escherichia coli* bacteria; and factor VIII (for haemophiliacs); and tissue plasminogen activator (tPA, for heart attack or stroke patients), both of which are produced in GM mammalian cells grown in laboratory culture. Furthermore, GM plants that produce "edible vaccines" are under development. Such plants, which are engineered to express antigens derived from microbes or parasites that infect the digestive tract, might someday offer a safe, cheap, and painless way to provide vaccines worldwide, without concern for the availability of refrigeration or sterile needles. In addition, novel DNA vaccines may be useful in the struggle to prevent diseases that have proved resistant to traditional vaccination approaches, including HIV/AIDS, tuberculosis, and cancer.

Genetic modification of insects has become an important area of research, especially in the struggle to prevent parasitic diseases. For example, GM mosquitoes have been developed that express a small protein called SM1, which blocks entry of the malaria parasite, *Plasmodium,* into the mosquito's gut. This results in the disruption of the parasite's life cycle and renders the mosquito malaria-resistant. Introduction of these GM mosquitoes into the wild may someday help eradicate transmission of the malaria parasite without widespread use of harmful chemicals such as DDT or disruption of the normal food chain.

Finally, genetic modification of humans, or so-called gene therapy, is becoming a treatment option for diseases ranging from rare metabolic disorders to cancer. Coupling stem cell technology with recombinant DNA methods may someday allow stem cells derived from a patient to be modified in the laboratory to introduce a desired gene. For example, a normal beta-globin gene may be introduced into the DNA of bone marrow-derived haematopoietic stem cells from a patient with sickle cell anaemia. Introduction of these GM cells into the patient could cure the disease without the need for a matched donor.

Role of GMOs in Environmental Management

Another application of GMOs is in the management of environmental issues. For example, some bacteria can produce biodegradable plastics, and the transfer of this ability to microbes that can be easily grown in the laboratory may enable the wide-scale "greening" of the plastics industry. Zeneca, a British company, has already developed a microbially produced biodegradable plastic called Biopol. This is made using a GM bacterium, *Ralstonia eutropha*, to convert glucose and a variety of organic acids into a flexible polymer. GMOs endowed with the bacterially encoded ability to metabolize oil and heavy metals may provide efficient bioremediation strategies.

Genetic modification technologies may also help save endangered species such as the giant panda, whose genome is being sequenced in an international effort led by the Beijing Genomics Institute at Shenzhen. Genetic studies of the panda genome may provide insight into why pandas have such low rates of reproductive success in captivity. A likely set of genes to consider for future genetic modification, should the goals of panda conservation warrant it, is the major histocompatibility complex (MHC). The MHC genes play an important role in regulating immune function and also influence behaviours and physiological patterns associated with reproduction.

Socio-political Relevance of GMOs

While GMOs offer many benefits to society, the potential risks associated with them have fuelled controversy, especially in the food industry. Many sceptics warn about the dangers that GM crops may pose to human health. For example, genetic manipulation may potentially alter the allergenic properties of crops. However, the more established risk involves the potential spread of engineered crop genes to native flora and the possible evolution of insecticide-resistant "superbugs". In 1998 the European Union (EU) addressed such concerns by implementing strict GMO labelling laws and a moratorium on the growth and import of GM crops. In addition, the stance of the EU on GM crops has led to trade disputes with the United States, which, by comparison, has accepted GM foods very openly. Other countries, such as Canada, China, Argentina, and Australia, also have open policies on GM foods, but some African states have rejected international food aid containing GM crops.

The use of GMOs in medicine and research has produced a debate that is more philosophical in nature. For example, while genetic researchers believe they are working to cure disease and ameliorate suffering, many people worry that current gene therapy approaches may one day be applied to produce "designer" children or to lengthen the natural human lifespan. Similar to many other technologies, gene therapy and the production and application of GMOs can be used to address and resolve complicated scientific, medical, and environmental issues, but they must be used wisely.

GLOSSARY

A

allele Any one of two or more genes that may occur alternatively at a given site (locus) on a chromosome.
allopatric Having separate territories.
anagenesis Genetic change in a lineage.
analogy/analogous Correspondence of features due to similarity of function but not related to common descent.
antigens Protein that stimulates the immune response.
autosomes Chromosomes that are not sex chromosomes.

B

bioinformatics Mathematical analysis of biological systems.

C

centromere Attachment point for the nuclear spindle fibres that move chromosomes during cell division
chromatography Technique for separating and identifying chemical compounds.
cladogenesis Branching of lineage from common ancestors.
clone Group of individual cells or organisms descended from one progenitor; in biotechnology, multiple copies of a single fragment of DNA such as a gene.
codon Three adjacent nucleotides; the sequence of a codon dictates amino acid production.

complementation A form of interaction between non-allelic genes.
convergence similarity of features that evolved independently through similar functionality.

D

diploid Having two sets of chromosomes.
dominance Greater influence by one of a pair of genes (alleles) that affect the same inherited character.

E

electrophoresis A means of separating macromolecules based on the differences in their electric charge.
enzyme A protein with a specific catalytic function.
epistatic gene A gene that masks the phenotypic effect of another gene, such as albinism.
eukaryotes Cells or organisms that have a nucleus.

F

fitness Relative probability that a hereditary characteristic will be reproduced.

G

gamete Sex, or reproductive, cell containing only one set of dissimilar chromosomes, or half of the genetic material necessary to form a complete organism (i.e. haploid).
gene Functional unit of heritable material that is found within all living cells.
gene therapy Insertion of genes encoding a needed protein into a patient's body or cells.
genetic equilibrium Stable set of gene frequencies in a population.
genetically modified organism (GMO) Organism carrying a segment of DNA (transgene) from another organism.
genome The complete genetic complement of an organism.
genomics Study of the structure, function, and evolutionary comparison of whole genomes.
genotype Set of genes that an offspring inherits from both parents, a combination of the genetic material of each.

H

haploid Having only one set of chromosomes.
heterozygous Having differing paired alleles for a trait.
homology/homologous Similarity of the structure, physiology, or

development of different species of organisms based upon their descent from a common evolutionary ancestor.
homozygous Having identical paired alleles for a trait.
horizontal transmission Transmission of a gene directly from one species to another.

I

introns Regions of DNA that are not translated into protein.

K

karyotype Physical appearance of the chromosome.

L

locus The site of a particular gene on a chromosome.

M

meiosis Division of a germ cell involving two fissions of the nucleus and giving rise to four gametes or sex cells, each possessing half the number of chromosomes of the original cell.
mitochondria Organelles involved in energy metabolism.
mitosis A process of cell duplication, or reproduction, during which one cell gives rise to two genetically identical daughter cells.

N

nature–nurture Diversity between individuals caused by hereditary versus environmental influences.
nucleotides Any member of a class of organic compounds in which the molecular structure comprises a nitrogen-containing unit (base) linked to a sugar and a phosphate group.

O

organelle A specialized structure within a cell.

P

parallelism Evolution of similar features in organisms without common ancestors.
paramutation Non-Mendelian inheritance of a genetic variation arising from interactions between pairs of identical alleles.
pathogen Disease-causing microorganism.
phenotype Organism's outward appearance and characteristics.

phylogeny Evolutionary history of an organism.
plasmids Extragenomic circular DNA elements.
polygenes Several or many genes affecting the phenotype.
polymerase chain reaction A method for rapidly detecting and amplifying a specific DNA sequence.
polymorphism Presence of two or more distinct hereditary forms associated with a gene.
polyploidy Multiplication of chromosome sets.
prokaryotes Cells or organisms lacking internal membranes, i.e. bacteria.
purine Any of a class of organic compounds of the heterocyclic series characterized by a two-ringed structure composed of carbon and nitrogen atoms.
pyrimidine Any of a class of organic compounds of the heterocyclic series characterized by a ring structure composed of four carbon atoms and two nitrogen atoms.

R

recessiveness The failure of one of a pair of genes (alleles) present in an individual to express itself in an observable manner because of the greater influence, or dominance, of its opposite-acting partner.
recombinant DNA DNA from two different sources "spliced" together by restriction enzymes, used in the cloning process.
reproductive isolating mechanism (RIM) Property of organisms that prevent interbreeding, such as geographic separation.
restriction enzyme A bacterial protein that cleaves DNA at specific sites along the molecule.

S

speciation Process by which new species are formed.
stem cell Undifferentiated cell that has the potential to become a specialized cell.
substrate The substance on which an enzyme acts.
sympatric Having the same territory.
synteny Having chromosome sequences in common.

T

transgene Gene from one organism inserted into a chromosome of another.

W

wild type Most common genotype found in natural populations.

INDEX

47,XYY syndrome 239, 241

ABO blood groups 89, 222, 346
acetylation 40
Achillea, environmental experiments 97
achondroplasia 151, 242, 243–4
acquired characteristics, inheritance 57, 65
activators 23
adaptation
  effect of mutations 139, 140
  of forelimbs 198
  and natural selection 69, 79, 148
adenine 4, 5
aflatoxin 258
ageing 42, 268–72
agriculture 48–9, 56, 286
  see also animal breeding; crop species; plant breeding
Albertus Magnus 64
albinism 45, 86, 89–90, 151, 245
alcohol 261
alkaloids 258
alkaptonuria 44, 60, 87
alleles
  elimination 150–1
  frequency 131
  interactions 88–9
  multiple 88–9
  mutant expression 18–19
  neutral 212
  segregation 83
allopolyploidy 114, 183–4
alphafetoprotein (AFP) 265–6, 338
altruism, reciprocal 163–4
Ames test 259
amino acids
  20 different types 9
  folding of chains 13, 16
  sequences yield genetic information 75
  substitution 136
amniocentesis 48, 266, 339
amplification, in recombinant DNA technology 45
amyotrophic lateral sclerosis (ALS) 264
anagenesis 188–9, 204, 205
analogy 190
Anaximander 63
anencephaly 265–6
aneuploidy 114–15, 266, 271
Angelman syndrome 41, 249, 251
animal behaviour 39, 77–8, 164–5, 170, 171, 253
animal breeding 46, 48–9, 53, 92, 128, 132, 133, 275–82
animal models
  in evaluation of populations 281–2
  for human disease 353, 362
animals
  cloning 358, 360
  polyploidy 182
  symmetry 199–200
anoles, ecotypes 118
Anomalocaris canadensis 202
anthropometric measurements 347
antibiotics, manufacture 49
antibodies 46, 218, 219–20
anticipation, in non-Mendelian inheritance 250
anticodon 16
antigens 46, 218, 219, 320
apomixis 114
apple maggot 118
Aquinas, Thomas 64, 67
argument from design 67–8
Aristotle 56
artificial insemination 48, 283, 284
artificial selection 132–3, 157
Augustine 63
Australopithecus 122, 158
autopolyploidy 183, 184
autosomal dominant inheritance 241–5, 254, 333–4
autosomal recessive inheritance 245–6, 267
autosomes 107
Avery, Oswald T. 54, 60
Axel, Richard 225

B lymphocytes 218, 220
Babylonians, animal breeding 53
Bacillus thuringiensis 361
back mutation 17, 21, 144
back-cross method, in plant breeding 296–7
bacteria
  chromosome mapping 106
  genetically modified 364
bacterial artificial chromosomes (BACs) 308
bacterial genetics 35
bacterial toxins 258–9
bacteriocins 259
bacteriophages 54, 106
Bakewell, Robert 275
Barr body 240
Bateson, William 73, 86

Bayes's theorem 335–8
Beadle, George 60
behavioral genetics 39, 253
bent grass *Agrostis* 160
Benzer, Seymour 61
Berg, Paul 62
Binet, Alfred 347
bioassays, in nanomedicine 325–7
biochemical methods 43–4, 75, 267–8
biogeography 77
bioinformatics 29, 38, 47, 48, 305, 322
biological determinism 233, 348
biometricians 73
bioremediation 364
biotechnology 37, 49, 303–6
birth defects 92, 96, 234, 235, 335
blending inheritance 70, 87–8
blood, human genetics 222–5
blood groups 46, 88, 89, 222–3, 346
blood theory of inheritance 56–7, 91
body plans 200
bone marrow 355–6
bovine somatotropin (BST) 306
Boyer, Herbert W. 303, 304
breast cancer, familial 31–2, 255
breeding
  analysis 48
  experimental 42–3
  selective 92, 275, 276
  see also animal breeding; plant breeding
Brenner, Sydney 61
brewing industry 49
Bridges, Calvin 69, 109
Bridgewater Treatises 67
Buck, Linda 225
Buffon, comte de (Georges-Louis Leclerc) 64
bulk-population method, in plant breeding 296
Burkitt lymphoma 257

cancer
  epigenetic changes 41–2
  familial 253–4
  genetic aspects 19, 20, 253–7
  multifactorial 252, 254
  prenatal diagnosis 266
  protecting against 219, 222
  successive mutations 256
cap 12
carcinogenicity, testing 259–60
beta-carotene, in golden rice 361–2
carrier testing 333, 341, 342
cattle
  carcass quality 278
  cross-breeding 285
  evaluation 280–1
  milk production 132, 276, 279–80
  progeny testing 283–4
  reproductive techniques 283
  Shorthorn, lack of dominance 88
cell
  membranes and organelles 14–15
  specialization 41
cell division, copying errors 17
cell fusion experiments 358
cellular immunity 218–19, 221–2
centromere 100, 112
cereal grains
  domestication 286
  see also crop species
Chargaff, Erwin 5
Chase, Martha 54
cheetah, lack of allelic variation 148
chickens, egg production 132
chimpanzee
  behaviour 164
  genomic information 30
  relation to humans 76, 78, 121, 127
chlorophyll, synthesis 55
chloroplast DNA (cpDNA) 25
chorionic villus sampling 48, 266, 339–40
Christianity, traditional beliefs 63–4, 67
chromatids 100, 101
chromatography 44
chromosomal abnormalities 110–15
  diagnosis 43, 266
  numerical 236–7
  structural 237–8
chromosomal disorders 235–41, 334
chromosomal mutations 140–1, 181–2
chromosome mapping 106
chromosome number 48–9, 100–1, 113–15, 141
chromosomes
  behaviour during cell division 100–3
  contain genes 9, 10, 12, 59, 104–6
  crossing-over 59, 102, 321
  deletion 111, 141
  diploid/haploid number 101
  duplication 111–12, 141
  early studies 99–100, 321
  homologous 110, 140–1
  human 100, 103, 106, 115–17
  inversion 112, 141
  karyotyping 265, 266
  microscopic study 36
  possible combinations 103
  structural changes 110–11
  translocation 112–13, 141
chronic granulomatous disease (CGD) 264
chronic myelogenous leukaemia 256
cicadas, temporal isolation 169
cigarette smoke *see* smoking
cladogenesis 188–90, 204, 205
Clarkia spp., quantum speciation 181–2
cleft lip and palate 253
clinical genetics 38–9
clones/cloning 48, 307, 312–13, 358, 360
co-dominance 88, 89
coat colour, in mammals 89
codons 13, 61, 136, 137
Cohen, Stanley N. 304
colchicine 48
Collins, Francis 33
colonization 156, 176, 178
colorectal cancer, familial 32, 255–6, 342, 345
colour blindness, sex-linked inheritance 108
colouration, protective 160
combustion products, mutagenic 260–1
comparative genomic hybridization (CGH) 29
competition, and natural selection 129
complementary DNA (cDNA) library 311–12
complementation 90–1
Condorcet, marquis de 64
congenital defects *see* birth defects
convergence 191
Copernican revolution 66–7
copying errors, in cell divisions 17
corn (maize)
  chromosome mapping 106
  chromosome number 101
  high-lysine varieties 289

hybrid varieties 297, 299–300
linkage groups 105
oil content 132
protein content 132
Correns, Carl Erich 81
cosmids 308
cotton *Gossypium* spp.
genetically modified 361–2
hybrid breakdown 173
cows *see* cattle
CpG islands 313
Creighton, Harriet 60
cri-du-chat syndrome 111, 238
Crick, Francis 5, 33, 60, 61, 75
crop species
disease resistance 301
extending area of production 288
genetically modified 361–2
pollination 291–2
suitable for mechanized agriculture 288–9
transgenic lines 49
cross-breeding 284–5
cross-pollination 82, 291, 292
crossing-over 59, 102, 321
cystic fibrosis 245, 267, 337–8, 344, 345
cystinuria 44
cytochrome c 119, 120, 127, 204–5, 206, 207–8
cytogenetics 36–7, 43
cytology 36
cytoplasmic DNA *see* DNA, extranuclear
cytosine 4, 5
cytosterility 298, 300
cytotoxic killer cells 219, 222

Darwin, Charles 58, 65–9, 74, 95, 121
theory of natural selection 69–72, 128–9
Darwin, Erasmus 64
Darwin's finches *see* Galapagos finches
DDT, resistance to 157
deafness
congenital, screening for 343
hereditary 91
deletions 237
Dendrobium orchid 169
deoxyribonucleic acid *see* DNA
deoxyribose sugar 4
depression, hereditary components 38
depurination 20
descent with modification 190, 198
diabetes 252, 345
diagnostics, novel 317–18, 325–7
Diamond v. *Chakrabarty* 304
Diderot, Denis 64
dietary management, in treatment of genetic disease 345
dinosaurs, extinction 197
dioxins, carcinogenic 259
disease
epigenetic contributions 41–2
resistance, in plants 301–2
susceptibility to 96, 98
see also genetic disease
disomy, uniparental 251
DNA
action of ultraviolet light 262
ageing 270–1
analysis, in genetic testing 344
as carrier of genetic information 8, 53–4, 203–5
chemical modifications 39–40
cloning 76, 307–13
composition 4–6
discovery 53
double helix 5, 6, 60, 61
extranuclear 25
homology 30, 313
human *see* human genome
isomerization 93
mutations 17
origin of replication 8
in preserved tissues 27
repair 254, 256, 263
repetitive 24
replication 7, 8, 10, 61
separation entropy 326
sequencing 29, 62, 76, 206–7, 313–14, 321, 322
spontaneous isomerization 20
structure 4–6, 75
synthesis 10
transcription 8, 10
see also recombinant DNA
DNA fingerprinting 318–19
DNA library 311–12
DNA ligase 310
DNA phylogeny 119–21
DNA polymerase 8
DNA probes 266–7
DNA topoisomerases 8
DNA vaccines 320
DNA-assisted approaches, in nanofabrication 328
Dobzhansky, Theodosius 74, 79
dogs, hip dysplasia 282
Dolly the sheep 357, 360
domesticated animals
colour patterns 89
see also animal breeding
dominance hierarchy 162
dominance inheritance, dominance, not always clear-cut 87–8
double helix *see* DNA, double helix
Down syndrome 43, 114, 236, 237, 266, 334, 339
Drosophila
chromosome mapping 106
chromosome number 101
early studies 59, 61
ethological (behavioral) isolation 170–1
gene linkage 104–6
gene mapping 59, 61
genome 77
Hox genes 202
as model organism 38
native Hawaiian species 181
pleiotropic white gene 87
rate of evolution 131–2
sex-linked inheritance 107–8, 109
sexual selection 162
speciation 177–8, 185–7
drug resistance 140
Duchenne muscular dystrophy 248, 335–7

ear, mammalian, evolution 199
earth sciences, conceptual revolution 77
ecology 77–8
ecotypes 118
egg donation, legal and ethical problems 357
Ehlers–Danlos syndrome 345
electromagnetic radiation, mutagenic 138
electrophoresis 44, 75
embryonic development 199–202
Empedocles 63
endoplasmic reticulum 14
enhancers 23
Enlightenment 64
environment
changing 149, 152, 156, 159
heterogeneity 159

human influences 157
impact on variation 276
modifies gene expression 226–7
new *see* colonization
environmental factors
and epigenetic changes 41–2
experiments 96–7
and fitness 152–3
genetic damage from 257–64
heredity 94–8
mutagenic 60
environmental management, use of GMOs 364
enzymatic defects, biochemical tests 267–8
enzymes
assays 75
genetically engineered 305
mutations affecting active site 19
three-dimensional structure 16
epigenesis 94–5
epigenetics 39–42
epistasis 89–90
Epstein–Barr virus 258
Escherichia coli 7–8, 22, 44–5
ethical issues
genetically modified organisms (GMOs) 305–6, 359–60, 364–5
human egg donation 357
human embryonic stem cells 353–4, 358
human genetics 331–2
evolution
convergent and parallel 190–3
and embryonic development 199–202
of eukaryotes 198
functional shifts 199
as genetic function 128–41
molecular 76–7, 206–13
and natural selection 57–8, 128–9
rate 131–2, 167–8, 193–5, 208–9
see also molecular clock
synthetic theory 73–5
evolutionary relationships, and genetic similarity 3–4, 313–14, 324–5
evolutionary theory, history 63–79
evolutionary trees *see* phylogenies
evolutionary trends 158
exons 12
expression libraries 312
expressivity, variable 244
extinctions, mass vs background 196–9
eye colour, hereditary 227
eyes, evolution 198–9

familial adenomatous polyposis (FAP) 255–6
family history, and genetic disease 47, 333
feathers, evolution 198
fertilization, in humans 115
fetal malformations, prenatal diagnosis 338
fibrinopeptides, rapid evolution 208
fibroblast growth factor receptor 3 (FGFR3) gene 242
first-cousin marriages 155
Fisher, R.A. 73, 131
fitness
Darwinian (relative) 129
environmental factors 152–3
measuring 149–50
flower structure, may impede pollination 171–2
fluorescent in situ hybridization (FISH) 29, 267, 340
fluorescent tagging 43
Flynn, James 349
foods, genetically modified 361–2, 365
forelimbs, adaptations 198
forensic medicine, use of PCR technique 27
forward mutation 17, 144
fossil fuels, combustion products 260
fossil record 120, 122, 124, 125, 189, 193, 194, 199
founder principle 147
fragile-X syndrome 18, 248, 249, 250
frameshift mutations 18, 19, 137–8
Franklin, Rosalind 5, 60
frequency-dependent selection 152–4
fruit fly *see Drosophila*

gain-of-function mutation 256
galactosaemia 44, 226
Galapagos finches 118, 179–80
Galton, Francis 231
gamma globulins 224
Garrod, Archibald 60, 86–7
Gaucher syndrome 345
gene expression
environmental effects 226–7
regulation 21–4
variability 90
gene flow 145, 174
gene frequency 144–8
gene pool 130, 167
gene regulation 21–4, 62
gene therapy 37, 48, 316–17, 345, 363
Genentech 303, 304
genes
artificially induced changes 302
ascribing functions 314, 322–3
behavioral effects 39, 253
discovery 34
disruption 315
dominant 83–5
duplication 120, 192, 207, 325
environmental damage 257–64
epistatic vs hypostatic 89–90
evolutionary conservation 3, 119
finding 313, 322
horizontal transmission 121
human *see* human genome
identification by microarray genomic analysis 323–4
imprinted 41, 249, 251–2
interactions 55, 89–94
knockout 315, 317
linear structure 61
located in chromosomes 9, 10, 12, 59, 104–6
molecular evolution 206–7
multiple alleles 88–9
mutations 17, 135–40
as physical basis of heredity 99–100
regulation *see* gene regulation
regulatory region 10
transcription 9, 10–12
word first used in 1909 55, 82
genetic ancestry projects 31
genetic code
discovery 60–1
expression 8–10
structure 12–14, 136–7
uniformity 14, 61, 125, 126
genetic conditioning 95
genetic correlation, in animal breeding 279–80

genetic counselling 38–9, 317, 333–8
genetic disease
animal models 353, 362
calculating risks 333–8
chemical testing 44
chromosomal abnormalities 235–41
diagnosis 31–2, 47–8, 341–4
importance 234
imprinted gene mutations 249, 251–2
management 332
Mendelian inheritance 333–4
multifactorial inheritance 252–3
mutations 18, 151
non-Mendelian inheritance 249–52
single-gene 38, 249–52, 333–4
treatment 344–5
X-linked 334
genetic distance 185, 188
genetic drift 129, 145–8, 211–13
genetic engineering 37, 133
see also genetically modified organisms (GMOs); recombinant DNA
genetic fingerprinting 27
genetic gain, in animal breeding 280–1, 284
genetic homeostasis 156
genetic information systems, ageing 270–2
genetic lineages, branching 76, 78
genetic linkage 59–60, 104–5, 141
genetic markers, in prenatal diagnosis 318
genetic monomorphism 159
genetic polymorphism *see* polymorphism
genetic testing 265–8, 341–4
genetic variation
biochemical study 75
and differential reproduction 128–9, 148
and directional selection 157
and evolution 130–2
in humans 31, 134
origin 135
in plants 287
in populations 31, 130–4
quantifying 132–4
and selective breeding 276–7
see also variation
genetic–environmental interactions 226–7, 231, 276
genetically modified organisms (GMOs)
in agriculture 42–3, 316, 360–2
in environmental management 364
ethical issues 305–6, 359–60, 364–5
in medicine and research 362–3, 365
potential risks 364–5
socio-political relevance 364–5
genetics
of antibody formation 219–20
biochemical techniques 43–4, 75
classical 34
definition 3
ethical issues 331–65
history 53–63
immunological techniques 46
impact on human affairs 4
industrial applications 49
mathematical techniques 46–7
medical applications 47–8
microbial 35
molecular techniques 37, 45–6, 60
and natural selection 72–3
physiological techniques 44–5
role in defining traits and health risks 31
theoretical 73–4
genomes
of eukaryotes 77
human 27–33, 48
genomic imbalance 110, 111
genomic information, comparison 30
genomic library 45, 311
genomic sequences, public databases 314, 322
genomics 37–8, 39, 44, 305, 321–5
functional 323
genotypes
constancy 80, 81
mixture 154
germ cells, embryonic 354
germ plasm theory 72, 95
giant panda, genome sequencing 364
Gilbert, Walter 62, 314
glycogen-storage disease 267
Golgi apparatus 15
gopher *Thomomys talpoides*, speciation 182, 188
gorilla, relation to humans 76, 78
gout 44
graft-versus-host disease 356
grains, hybrid 48
Greek philosophy 56–7, 63
Gregory of Nazianzus 63
Griffith, Frederick 53
guanine 4, 5

H-Y antigen 115
haemochromatosis, genetic testing 267
haemoglobin 44, 192–3, 207, 225
haemophilia 108, 248–9, 267, 345
Haldane, J.B.S. 73
haptoglobins 224
Hardy, G.H. 142
Hardy–Weinberg equilibrium 35–6, 47, 142–4
Hawaii, speciation 173–4, 180–1
hearing loss *see* deafness
heart disease 38, 252
heavy metal contamination 160
helicases 8
hepatitis B and C viruses 258
heredity
basic features 81–94
concept 80
environmental factors 94–8
influence on behaviour 39, 253
physical basis 99–100
heritability 95–8, 278–9
hermaphrodites 116, 182
herpes virus 258
Herrnstein, Richard 350
Hershey, Alfred D. 54
heterosis 151–2, 285, 297, 299
heterozygotes 83, 133–4
hip dysplasia, in dogs 282
Hippocrates 56
histones 40, 41–2, 93
Homo erectus 122, 167
Homo habilis 122
Homo sapiens (modern humans)
evolution 121–3, 167
origin and migration 30–1, 36, 122, 224
see also humans
homology 190–1, 192

homozygotes 83
homunculus 94
honeybees, social behaviour 164–5
horse, evolution 158, 188
host resistance and selection
house mouse *Mus musculus*
  chromosome mapping 106
  chromosome number 100
  genome 77
  linkage groups 105
Hox genes 200–2
human chromosomes
  mapping 106
  number 100, 103
  and sex determination 115–17
human embryonic stem cells 353–4
human evolution 30, 121–3
human genetics
  ethical issues 331–2
  research and applications 38–9
human genome 27–33, 48
Human Genome Project 32–3, 76–7
human immunodeficiency virus (HIV) 10, 258
human leucocyte antigens (HLA) 46, 221–2, 229
human population genetics 31, 36
humans
  aneuploidy/polyloidy 114
  behavioral disorders 253
  cloning 358
  cranial capacity 158, 188
  differentiation 117
  early interest in heredity 53, 56–7
  evolution 167–8
  genetic continuum 324–5
  genetic diseases *see* genetic disease
  genetic variation 31, 134
  inbreeding 155
  linkage groups 105
  longevity 270
  Mendelian inheritance 86–7
  population bottlenecks 148
  rare-mate advantage 154
  relation to apes 76, 78, 121, 122, 127
  sex ratio 116
  sex-linked inheritance 108
  skin colour 89
humoral immunity 218
Huntington disease 244, 249, 250, 267
Huxley, Julian 74
Huxley, T.H. 69
hybrid varieties, in plant breeding 297–8, 299–300
hybrid vigour *see* heterosis
hybridization 174–5, 294–7
hybridomas 220
hybrids 172, 173
hydrogen bonds 5, 9

immunogenetics 46, 218–22
immunoglobulins *see* antibodies
in vitro mutagenesis 314–15
inborn errors of metabolism 60, 86–7, 246
inbreeding 154–5, 285–6
individual characteristics, heritability 96
industrial chemicals, carcinogenic 259–60
industrial melanism 118, 140, 158
industrial processes, use of biotechnology 305
infant mortality 234
initiation complex 23
insecticides
  decreased use 360–1
  resistance 157, 365
  see also pesticides
insects
  genetic modification 363
  social 164–5
insertions 237
insulin, recombinant 304
intelligence
  genetic–environmental interaction 231, 349–50
  heritability 348–51
  measuring 347–9
  socioeconomic factors 348, 349, 350
interbreeding, and species concept 166–8
intestinal epithelium 355
intron splicing 12, 19, 23
introns 12, 312
ionizing radiation 20, 262–3
IQ tests 231, 347–8, 349–50
iron deficiency 362
islands, adaptive radiation 179
isolation, mechanisms 168–73

Jacob, Francois 61–2
Jeffreys, Alec 318
Johannsen, Wilhelm 55, 82
Judaism, traditional beliefs 63

karyotyping 266, 334, 344
keratinocytes 355
Khorana, Har Gobind 61
Kimura, Motoo 76
kin selection 163–5
Klineberg, Otto 348
Klinefelter syndrome 43, 115, 239, 240, 266

lac repressor 22
lactose 22, 226
Lamarck, chevalier de (Jean-Baptiste de Monet) 57, 65, 72
lambda phage 308
lamp shell *Lingula*, living fossil 194, 195
land varieties of plants 292, 294
leader sequences 24
Leber hereditary optic neuropathy 249
Leighton, E.A. 282
leptotene 101
Lesch–Nyhan syndrome 248
leukaemia inhibitory factor (LIF) 352, 353, 354
life, origin 10, 135
lineage, divergence 119–20
linkage groups 104–5
Linnaeus, Carolus 65
liver cancer 258, 259
living fossils 194
locus 130
longevity, inheritance 269–70
lung cancer 20, 260–1
Lyon, Mary 240

McCarty, Maclyn 60
McClintock, Barbara 60
Macleod, Colin M. 60
macroevolution 119
major histocompatibility complex (MHC) 221, 364
malaria parasite *Plasmodium* 77, 152, 363
males, size and aggressiveness 161–2
Marfan syndrome 90, 244, 345
marine animals, gametic isolation 172
marsupials 191, 206
mass selection 280, 292–3, 299
maternal age, and chromosomal abnormalities 236
maternal inheritance, of mitochondrial DNA 250
mathematical methods, in genetics 35–6, 46–7, 73–4

mating
  assortative (selective) 142–4, 154–5
  and courtship behaviour 170, 171
  in plant matings 291–2
  random 142
Maupertuis, Pierre-Louis Moreau de 64
Maxam, Allan M. 62, 314
Mayr, Ernst 74
medical genetics 38, 217
medicine, genetic techniques 27, 47–8, 305
meiosis 59, 99, 100–3, 110, 236, 321
melanin 8, 45, 139
Mendel, Gregor 34, 55, 57, 81
  experiments 58, 71, 81–5
Mendelian genetics 57, 71, 81–8
Mendelian inheritance, of genetic disease 333–4
Mendel's laws 83–5, 86
  discovery and rediscovery 59, 71–2, 81–5
  and meiosis 102–3
  universality 59, 85–7
Meselson, Matthew 7, 61
messenger RNA 11, 12, 13, 23, 136
metabolic pathways, uniformity 125–6
methionine 13
methylation, of histones 40, 41–2, 93
microarray genomic analysis 323–4
microarray hybridization analysis 344
microevolution 118–21
microfluidic systems 326
microorganisms
  as model organisms 35
  in pharmaceutical industry 49
microRNA 94
microsatellite DNA 24
Miescher, Johann Friedrich 53
milk production *see* cattle, milk production
miniaturization 327
minisatellite DNA 24, 318
Mirabilis jalapa, hybrids 87–8
miscarriage (spontaneous abortion) 234, 235–6, 237, 238, 334
missense mutations 17, 20
mitochondrial DNA 14, 25, 249–50
mitosis 99, 100
MN blood groups 88, 130, 222, 346
mole rat *Spalax ebrenbergi*, speciation 182, 188
molecular biology
  in evolutionary studies 75, 125–8
  uniformity 125
molecular clock 76, 120, 209–11, 212
monoclonal antibodies 305
Monod, Jacques 61–2
Morgan, Thomas Hunt 59, 104–5, 107
morphological variants, selection 288
morphology, protective 160
mosaicism 239, 334
mosquitoes
  Anopheles gambiae, genome 77
  Anopheles maculipennis reproductive isolation 169
  genetic modification 363
motor neuron disease 264
mouse
  genetic modification 352–3
  see also house mouse
Müller, Hermann Joseph 69–70
Mullis, Kary B. 26, 63
Murray, Charles 350
mustard weed *Arabidopsis thaliana*, genome 77
mutagenicity, testing 259
mutagens 20, 60, 138, 302
mutation–selection equilibrium 145
mutational analysis, in vitro 314–15
mutational pressure 145
mutationism 73
mutations
  and adaptation 139, 140
  artificial 70
  continuous process 139–40
  dominant 18
  and evolution 135–41
  expression 18–19
  full expansion 250
  and gene frequency 144–5
  impact 18–19, 138–9
  imprinted genes 249, 251–2
  leaky 19
  mechanisms 17–20
  for neutral alleles 212
  random 17
  rates 139, 151
  recessive 18–19
  repair 20–1
  second-site 21
  somatic 19
  spontaneous 20, 129, 138, 334
  within introns 19
myelomas (plasma cell tumours), in immunogenetic studies 220

naked DNA therapy 320
nanofabrication 327–8
nanotechnology 305, 325–8
Nathans, Daniel 62
natural selection
  and adaptation 69, 79, 148
  Darwin's theory 68–9
  definition 128–9
  and diversity 149
  and gene frequencies 150–4
  organizational effects 149
  in populations 148–65
  as process of genetic change 148–54
nature–nurture interaction
  investigation 96–7
  medical implications 233
  twin studies 229–30
  see also genetic–environmental interactions
nematode worm *Caenorhabditis elegans*, genome 77
neo-Darwinism 72
neo-Lamarckism 72
neural tube closure defects, prenatal diagnosis 338–9
neurodegenerative disorders 249
neurofibromatosis 90, 244
Neurospora crassa 60, 106
neutrality theory of molecular evolution 76, 211–13
newborn babies, screening 47, 343
Nirenberg, Marshall 61
non-Mendelian inheritance 92–4, 249–52
non-penetrance 90
nonsense mutations 17, 19
normal distribution 155
nucleic acids 4
nuclein 53
nucleotide sequence, alteration 135–9
nucleotides
  chemistry 4–5, 6
  polymerization 7
  sequencing 62, 63
null mutation 19

olfactory receptors 225–6
oligonucleotide-directed mutagenesis 315
oncogenes 254, 256, 258
open reading frames 313
operator region 22
operon model of gene regulation 22
organ transplantation, genetic aspects 219, 221
organisms
  common ancestry 125–7
  comparative studies of living forms 189–90, 203–4
  diversity 196, 197–8
  evolutionary relatedness 3, 14
  patentable 304
  structure and function 8
On the Origin of Species by Means of Natural Selection 66, 67–8, 166–7
osteogenesis imperfecta 244
outbreeding 154–5
overdominance 151–2
oxygen, molecular 263–4
ozone layer 261–2

pachytene 101
Paley, William 67–8
pangenesis 56, 58, 70, 95
papilloma viruses 258
parthenogenesis 271–2
patenting of organisms 304
paternity testing 46
Pearson, Karl 73
pedigree procedure, in plant breeding 295–6
pedigrees 242, 246, 247, 248, 282, 333, 336
penetrance 245
peppered moth, industrial melanism 118, 158
percutaneous umbilical blood sampling (PUBS) 339, 340
pest resistance, goal of plant breeding 288
pesticides, resistance 140, 157
pharmaceutical industry, genetic applications 49
pharmaceuticals
  designer drugs 48
  genetically engineered 304–5, 319, 362–3
phenotype, variation 80–1
phenylalanine 13, 16, 47
phenylketonuria (PKU)
  biochemical testing 44, 246
  dietary management 345
  inborn error of metabolism 246
  recessive inheritance 145, 150, 151
  screening 47, 267, 343
  single-gene disease 38
  symptoms 246
pheromones 171, 226
Philadelphia chromosome 256
phosphorylation 40
photolyase 262
photosensitization 262
phylogenetic tree, molecular 120
phylogenies 122, 189–90, 193, 205–6, 209–10
pigs, cross-breeding 285
plant breeding
  and artificial selection 132
  cross-pollinators 298–301
  goals 287–9
  history 128, 286–7
  self-pollinators 292–8
  synthetic varieties 300–1
  techniques 48–9, 92
plants
  disease resistance 301–2
  evaluation 289–90
  mating systems 291–2
  nutritional quality 289
  ornamental 289
  pest-resistant 48
  polyploidy 114, 182–4
  toxic compounds 258
  transgenic *see* genetically modified organisms (GMOs)
plasmids 25, 62, 307–8
plastics, biodegradable 364
pleiotropy 87
Plomin, Robert 349
pluripotency, induced 358–9
point mutations 17–18
pollination 291–2
poly(A) tail 12
polycystic kidney disease, genetic testing 267
polygenic inheritance 91–2, 96
polymerase chain reaction (PCR) 25–7, 29, 62–3, 76, 319
polymorphisms 118, 153, 159, 224, 225
polynucleotide 7
polyploidy 113–14, 141, 182–4
population genetics 35–6, 47
populations
  bottlenecks 147–8
  effective size 146
  genetic variation 31, 130–4
  geographic separation 176
  screening 342
porphyria 267
position effect 112
post-translational modification 23–4
potato, chromosome number 101
Prader–Willi syndrome 41, 249, 251, 267
preformation 57
preformism 94–5
pregnancy
  elective termination 341
  see also miscarriage
preimplantation testing 339, 340
prenatal diagnosis 43, 44, 47, 265–6, 317–18, 338–41
probability, estimating 47, 335–8
progeny testing 283–4
promoter region 11, 19, 22
protein medications *see* pharmaceuticals
proteins
  amino acid sequence 9
  effect of mutations 18, 19, 21
  evolutionary rate 208
  genetic coding 9
  as informational macromolecules 203–5
  pharmaceutical, genetically engineered 304–5, 319, 362–3
  properties 13
  recombinant, advantages 304–5
  regulatory, binding 62
  structure and function 8–9, 16, 18, 19, 21
  synthesis *see* translation
proteomics 305
proto-oncogenes 19, 256
punctuated equilibrium model 195
Punnett, Reginald 86
pure-line selection, in plant breeding 293–4
purines 4
pyrimidines 4
  dimerization 262

quantum speciation *see* speciation

race
  and IQ testing 347–8
  unscientific concept 122–3, 231, 345–6

radiation, adaptive 179–81
radioactive tagging 43, 44
Rassoulzadegan, Minoo 92–3
rat *Rattus norvegicus*, genome 77
reading frame 18
rearrangements 237
recessive genes 83–5
recessiveness, not always clear-cut 87–8
recombinant DNA
  in biotechnology industry 49, 304
  and chromosome mapping 106
  composition 309
  in gene therapy 48
  in genetic testing 267
  in genetically modified organisms 360
  technology 45, 62–3, 133, 306–7
recombination 105–6, 129
red blood cells 28, 46
regenerative medicine 305
repeats, dispersed 24
replication, origin 8
reproduction, differential 128–9
reproductive isolating mechanisms 168–9, 170–1, 173–4, 175
reproductive techniques 283
restriction enzymes 62, 309
restriction fragment length polymorphisms 318
retinoblastoma 254–5
retroviruses 10, 258
Rett syndrome 41
reverse genetics 317, 323
reverse mutation *see* back mutation
reverse transcriptase 10, 311
reversion *see* back mutation
rhesus (Rh) system 222–3, 346
ribonucleic acid *see* RNA
ribonucleotides, polymerization 10
ribose sugar 4
ribosomal RNA (rRNA) 11, 14
ribosomes 14–16
rice
  genetically modified 361–2
  selective breeding 288
RNA
  and non-Mendelian inheritance 92–4
  nucleotide sequence 9
  in retroviruses 10
  structure and composition 4–6
  transcripts 10–11
RNA polymerase 10–12, 22

sage *Salvia* spp., mechanical isolation 171–2
Sanger, Frederick 62, 314
satellite DNA 24
schizophrenia 38, 232
screening, genetic 47, 267, 342
sea urchins *Strongylocentrotus* spp., gametic isolation 172
segregation 84, 85
selection
  accuracy 282
  in animal breeding 280
  directional 156–8
  diversifying 158–61
  in plant breeding 292–8
  sexual 161–2
  stabilizing 155–6
selection coefficient 150
self vs non-self 218
self-assembly, in nanofabrication 327–8
self-fertilization 144, 182
self-pollination 291, 292
serum proteins 223–4
serum testing, in prenatal diagnosis 338
sex chromosomes 106–8, 236
  anomalies 114–15, 238
sex determination 107, 108, 115–16, 236, 283
sex-linked inheritance 34, 106–9, 246–9
sexual dimorphism 161–2
sexual selection 154
sickle cell anaemia 150, 152, 225, 245, 267, 343–4, 345
Simpson, George Gaylord 74
single nucleotide polymorphisms (SNPs) 45–6
single-gene diseases 38, 249–52
site-directed mutagenesis 63
skin cancer 20, 21, 261–2
skin cells, renewal 28
skin colour 325
smell, genetic aspects 225–6
Smith, Hamilton Othanel 62
Smith, Michael 62–3
smoking, health risks 20, 260–1
sociobiology 78
somatic cell nuclear transfer 357–8
somatic mutation hypothesis 268–9, 270–1
sorghum 288, 297–8
speciation
  allopatric 176, 179
  cladogenetic function 188–9
  in *Drosophila* 177–8
  genetic differentiation 185–8
  geographic 176–8, 185–7
  model 173–6
  quantum 181–2, 188
  recent examples 118–19
  sympatric 181
species
  concept 166–8
  constancy and variation 80
  definition 166, 168
  endangered 364
  new 73, 74
  recognition signals 171
Spencer, Herbert 70
spina bifida 265
spontaneous generation 64
squash technique 36–7, 43
Stahl, Franklin W. 7, 61
staining, selective 43
start codon 13, 14
statistical methods 47, 335–8
Stebbins, George Ledyard 74
stem cells
  adult 354–7
  differentiation 41, 42
  embryonic 351–4
  epithelial 355
  haematopoietic 355–6
  in immune system 218
  induced pluripotency 358–9
  neural 356–7
  therapeutic possibilities 351
stop codons 13, 16, 136, 313
  erroneous 17–18, 21
Streptococcus pneumoniae, virulence 53–4
Sturtevant, Alfred Henry 59
superovulation 283
superoxide dismutase 264
survival of the fittest 70
Sutton, Walter S. 103
swallowtail butterfly *Papilio dardanus*, mimicry 160–1
Swanson, Robert A. 303
sweet potatoes, genetically modified 361–2
synaptonemal complex 110
synteny 121, 122

T lymphocytes 218–19, 221
tandem repeats 24
Taq polymerase 27
TATA box 12

Tatum, Edward 60
Tay–Sachs disease 38, 245, 265, 342
Terman, Lewis 347
termination sequences 12
tetraploidy *see* polyploidy
thalassaemia 225, 245, 265, 343, 345
therapeutic molecules *see* pharmaceuticals
Thermus aquaticus 27
thiamin, synthesis 44–5
thymine 4, 5, 44
tissue-culture techniques 43
tomato
  chromosome number 101
  genetically modified 306
traits
  dominant 34
  genetic–environmental interaction 276
  inheritance 81–5
  intermediate 34
  linkage 103–9
  polygenic 34
  qualitative 232, 289–90
  quantitative 91–2, 290
  recessive 34
  sex appeal 161–2
  term of convenience 87
  transmission 34
transcription 9, 10–12, 19, 23
transfer RNA (tRNA) 11, 15–16
transferrins 224
transformation, in DNA cloning 311
transforming factor 54
transgenic modification 37, 48, 49, 316
transit amplifying cells 354
translation 9–10, 13, 14–16, 23
  termination 13, 18
translocations 237
  balanced 334
transmembrane domains 314
transposons 24
triple screen, in prenatal diagnosis 338–9
triplet *see* codon
triplet repeat expansions 18, 250
triploidy *see* polyploidy
trisomies, human 114, 236, 237, 239, 240
tryptophan 13
Tschermak von Seysenegg, Erich 81
tuatara *Sphenodon punctatus*, living fossil 194, 195
tuberculosis, susceptibility 232
tumour suppressor genes 42, 254, 256
tumours *see* cancer
Turner syndrome 114–15, 116, 238–9, 266, 345
twin studies 227–33, 349
twins
  fraternal 227–8
  identical 28, 134, 228–9

ubiquitination 40
ultrasound imaging, in prenatal diagnosis 338
ultraviolet radiation 20–1, 261–2
uracil 4

vaccines, novel 320, 362–3
variable number tandem repeats (VNTRs) 24
variation
  additive 277
  allelic 118, 148
  dominance 277
  environmental impact 276
  epistatic 277
  individual 80
  morphological 132
  see also genetic variation
vectors 35, 62, 121, 307–8, 309
Venter, J. Craig 33
vinegar fly *see Drosophila*
virulence, genetic differences 53–4, 232
viruses 35, 62, 257–8, 308, 320
de Vries, Hugo 72–3, 81

Waddington, Conrad 39
Wallace, Alfred Russel 57–8, 70
Watson, James D. 5, 33, 60, 75
Weinberg, Wilhelm 142
Weismann, August 72, 95
wheat, hexaploid 114
whole-genome replacement 360
wild type, definition 17
Wilkins, Maurice 5, 60
Wright, Sewall 73

X chromosome 106–7
  inactivation (lyonization) 239–40
  see also sex chromosomes
X-linked, genetic disorders 334
X-ray diffraction methods 5
X-rays, mutagenic 60
xeroderma pigmentosum 20–1, 262

Y chromosome 106–7, 239, 247
  see also sex chromosomes
Yanofsky, Charles 61
yeast
  improved strains 49
  Saccharomyces cerevisiae, genome 77
yeast artificial chromosomes (YACs) 308
yield, increase 288

ENCYCLOPÆDIA
Britannica

Britannica

Since its birth in the Scottish Enlightenment, Britannica's commitment to [illegible] and popular scholarship has never wavered. In 2008, Britannica celebrated its 240th anniversary.

Throughout its history, owners and users of Encyclopædia Britannica have drawn upon it for knowledge, understanding, answers, and inspiration. In the Internet age, Britannica, the online encyclopedia, continues to deliver that fundamental requirement of reference, information, and educational publishing – confidence in the material we read in it.

Readers of Britannica Guides are invited to take a FREE trial of Britannica's huge online database. Visit

http://britannicaguides.com

to find out more about this title and others in the series.